Deflagration and Detonation Flame Arresters

Deflagration and Detonation Flame Arresters

Stanley S. Grossel
Process Safety and Design, Inc.

CENTER FOR
CHEMICAL PROCESS SAFETY
An AIChE Industry Technology Alliance

CENTER FOR CHEMICAL PROCESS SAFETY
of the
American Institute of Chemical Engineers
3 Park Avenue, New York, New York 10016-5991

Library of Congress Cataloging-in-Publication Data

ISBN 978-0-8169-0791-5
Publication number G-64

Contents

3

Overview of Deflagration and Detonation Prevention and Protection Practices

4

Overview of Combustion and Flame Propagation Phenomena Related to DDAs

5
Deflagration and Detonation Flame Arrester Technology

6
Installation in Process Systems

7
Inspection and Maintenance of Flame Arresters

8
Regulations, Codes, and Standards

9
Illustrative Examples, Calculations, and Guidelines for DDA Selection

10
Summary

Preface

The Center for Chemical Process Safety (CCPS) was established in 1985 by the American Institute of Chemical Engineers for the express purpose of assisting industry in avoiding or mitigating catastrophic chemical accidents. To achieve this goal, CCPS has focused its work on four areas:

- Establishing and publishing the latest scientific, engineering, and management practices for prevention and mitigation of incidents involving toxic, flammable, and/or reactive material
- Encouraging the use of such information by dissemination through publications, seminars, symposia, and continuing education programs for engineers
- Advancing the state of the art in engineering practices and technical management through research in prevention and mitigation of catastrophic events
- Developing and encouraging the use of undergraduate engineering curricula that will improve the safety knowledge and consciousness of engineers

Despite precautions, unwanted combustion can occur in process piping and vessels. This practical book offers safety guidelines for the design, installation, maintenance and inspection of flame arrester systems to provide protection against deflagrations and detonations propagating through process equipment, piping and especially vent manifold systems. The combustion and explosion phenomenon is discussed as it relates to selection, installation and maintenance of deflagration and detonation flame arresters. Other methods to prevent these propagating flames such as venting, pressure containment, oxidant reduction, combustible concentration reduction, deflagration suppression, and equipment and piping isolation are also briefly discussed. A number of worked examples are given to illustrate various aspects of the design of these systems.

Acknowledgments

This book could not have written without the assistance of many people and organizations that shared their expertise and experience with me, and I would like to acknowledge help from the following people:

The work on the project was supervised by the CCPS Engineering Practices Subcommittee whose members provided appreciable input from their own expertise and experience. The CCPS Subcommittee was chaired by R. Walz (ABB Lummus Global) and included (in alphabetical order): L. G. Britton (Union Carbide/Dow), C. A. Dafft (Rohm and Haas), H. L. Febo, Jr. (FM Global), R. P. Gale (Solutia), H. G. Gurry (Procter & Gamble), P. N. Lodal (Eastman Chemical), G. Myers (U.S. Department of Energy), J. L. Owen (duPont), G. A. Peters (Air Products & Chemicals), S. A. Rogers (Honeywell), W. A. Thornberg (Industrial Risk Insurers), and A. Torres (Eastman Kodak).

John A. Davenport was the CCPS staff liaison and was responsible for the overall administration of the project.

I would like to thank Dr. Geraint O. Thomas (Centre for Explosion Studies, Department of Physics, University of Wales, Aberystwyth, UK) for offering to write Chapter 4. He prepared the first draft, but due to a sabbatical in Japan and other research commitments, he was unable to write the revisions. However, he did make constructive comments and suggestions on the revisions that I wrote. In addition, I would like to thank the following other individuals who provided me with technical data and other assistance based on their knowledge of flame arresters and combustion science and technology: G. Binks (IMI Amal), R. Butler (Enardo). K. Chatrathi

(Fike), D. Crowl (Michigan Technological University), J. DePasquale (ESAB), F. Destro (Western Enterprises), D. Dickerman (Praxair), J. Gorman (Varec), R. Guile (Western Enterprises), W. Howard (Consultant), D. Kirby (Union Carbide/Dow), K. Lapp (Westech), D. Long (IMI Amal), V. Mendoza (NAO), F. Nichols (retired from ICI), E. Patenaude (Westech), T. Piotrowski (Protectoseal), D. Pritchard (HSE), N. Roussakis (HAD Combustion and Chemicals), A. Schneider (U.S. Coast Guard), R. Schwab (Consultant), R. Schwartz (John Zink), R. Shepherd (Tornado Flare Systems), V. Smolensky (NAO), W. Stevenson (Cv Technology), J. Straitz III (NAO), M. Wauben (SchuF USA), D. Wechsler (Union Carbide/Dow), and R. White (Southwest Research Institute).

CCPS also gratefully acknowledges the comments and suggestions submitted by the following peer reviewers:

Geoff Binks	*IMI Amal Limited*
Michael Davies	*Braunschweiger Flammenfilter GmbH*
Randy Freeman	*ABS Group Inc.*
Margaret Gregson	*Health and Safety Executive (UK)*
Ken Lapp	*Westech Industrial Ltd.*
Vadim Smolensky	*NAO Inc.*
Geraint O. Thomas	*University of Wales, UK*
Anthony Thompson	*Monsanto Company*
Matt Wauben	*SchuF (USA), Inc.*
Donald White, Jr.	*Flair*
Jan Windhorst	*Nova Chemicals, Ltd.*

Lastly, I would like to express my appreciation to Syl Turicchi (former manager) and Jack Weaver (Director) of the CCPS staff for their support and guidance.

Acronyms and Abbreviations

AIT	Autoignition Temperature
API	American Petroleum Institute
ASTM	American Society for Testing and Materials
ATEX	Atmospheres Explosibles (European Protection Standard)
BAM	Bundesanstalt fur Materialprufung
BSI	British Standards Institute
CART	Calculated Adiabatic Reaction Temperature
CFR	Congressional Federal Register
CGA	Compressed Gas Association
CEN	Comité Européean de Normalisation
CJ	Chapman Jouguet
CMR	Crimped Metal Ribbon
CSA	Canadian Standards Association
DDA	Deflagration and Detonation Arrester
DDT	Deflagration-to-Detonation Transition
DMA	Detonation Momentum Attenuator
ESD	Emergency Shutdown
FM	Factory Mutual
FMRC	Factory Mutual Research Corporation
ft/sec	feet per second
HSE	Health and Safety Executive (UK)

IChemE	Institution of Chemical Engineers (UK)
IEC	International Electrotechnical Commission
IMO	International Maritime Organization
IR	Infrared
LEL	Lower Explosive Limit
LFL	Lower Flammable Limit
LOC	Limiting Oxidant Concentration
MESG	Maximum Experimental Safe Gap
MIC	Minimum Igniting Current
MIE	Minimum Ignition Energy
MOC	Maximum Oxidant Concentration
m/s	meters per second
NAS	National Academy of Sciences
NEC	National Electrical Code
NFPA	National Fire Protection Association
NMAB	National Materials Advisory Board
OSHA	Occupational Safety and Health Administration
PTB	Physikalische-Technische Bundesanstalt
RED	Restricted End Deflagration
UEL	Upper Explosive Limit
UFL	Upper Flammable Limit
UK	United Kingdom
UL	Underwriters Laboratories
UV	Ultraviolet
USCG	United States Coast Guard
VOC	Volatile Organic Compounds

1

Introduction

1.1. Intended Audience

This "concept book" is intended for use by chemical engineers and other technical personnel involved in the design, operation, and maintenance of facilities and equipment where deflagration and detonation arresters (DDAs) may be required. These people are usually technically competent individuals who are aware of, but not experts in, combustion phenomena. The facilities where such devices may be needed include chemical plants, petrochemical plants, petroleum refineries, pharmaceutical plants, specialty chemical plants, storage tank farms, loading and unloading facilities, and pipelines.

This book will also be of use to process hazard analysis (PHA) team members and process safety and loss prevention specialists.

1.2. Why This Book Was Written

There is a need in many chemical processes for protection against propagation of unwanted combustion phenomena such as deflagrations and detonations (including decomposition flames) in process equipment, piping, and especially vent manifold systems (vapor collection systems).

There are different ways, both passive and active, to provide this desired protection against deflagrations and detonations. Methods include DDAs, venting, pressure containment, oxidant concentration reduction (inerting and fuel enrichment), combustibles concentration reduction (ventilation or air dilution), deflagration suppression, and equipment and piping isolation. These are discussed in more detail in Chapter 3.

This book makes reference to flame arresters, deflagration flame arresters, and detonation flame arresters. "Flame arresters" is the generic term for both deflagration and detonation flame arresters. Deflagration flame arresters are used when a flame only propagates at subsonic velocity, whereas detonation arresters are used when a flame can propagate at all velocities including supersonic velocities.

One of the major reasons that this book was written is that nonspecialist chemical engineers know little about DDAs. Although DDAs have been specified and installed for many years, quite often they have failed because the wrong type of flame arrester was specified, or it was improperly installed, or inadequate inspection and maintenance were provided.

It is intended that this book will foster effective understanding, application and operation of DDAs by providing current knowledge on their principles of operation, selection, installation, and maintenance methods.

1.3. What Is Covered in This Book

This book covers many aspects of DDA design, selection, specification, installation, and maintenance. It explains how various types of flame arresters differ, how they are constructed, and how they work. It also describes when a flame arrester is an effective solution for mitigation of deflagrations and detonations, and other means of protection (e.g., oxidant concentration reduction) that may be used. It also briefly covers some aspects of dust deflagration protection.

Chapter 2 is a general discussion of the historical development of DDAs, and an overview of the applicable standards and codes is presented.

An overview of various prevention and protection methods against deflagrations and detonations is presented in Chapter 3.

Chapter 4 presents an overview of combustion and explosion phenomena as this is vital to the understanding of the conditions under which a DDA must function.

Chapter 5 is a comprehensive discussion of DDA technology, covering the various types of DDAs used in the chemical process industries, as well as selection and design considerations and criteria. Detailed information is presented on how to select a DDA for various operating conditions and applications (e.g., deflagration versus detonation conditions, end-of-line versus inline, vent manifold/vapor recovery systems). During the description and discussion of various types of DDAs, some application examples are presented.

Subsequent chapters cover installation considerations (Chapter 6), inspection and maintenance practices (Chapter 7), regulations, standards, and codes, including certification test protocols (Chapter 8), and some illustrative examples (Chapter 9). Chapter 10 provides a summary of the present state-of-the-art and what other information and research is needed, followed by appendixes, a glossary, and suggested additional reading.

This book does not provide specific recommendations for maritime operations (e.g., ship and barge loading and unloading). The requirements for these are covered in the U.S. Coast Guard regulations, which are outlined in Section 2.3.1 and in Chapter 8.

1.4. What the Reader Should Learn from This Book

After reading this book the reader should

1. be aware that it is not possible to design flame arresters from basic theory;
2. be more conversant with deflagration and detonation phenomena in process equipment and vent manifold systems;
3. know when a flame arrester is an effective solution for combustion hazards;
4. be able to select an appropriate flame arrester and have it properly installed;
5. know what needs to be done to keep a DDA functional;
6. be able to work with and ask the proper questions of "experts" and manufacturers.

1.5. Units of Measure

The equations given in subsequent chapters are presented as they appear in the original reference source. Some may have mixed units (English and metric) and, therefore, the numerical constants are not dimensionless.

2

History and State-of-the Art

2.1. Historical Development of Flame Arresters

The forerunner of the present-day flame arrester is the miner's safety lamp. In the early 1800s candles and oil lamps were used in coal mines and were responsible for many disastrous explosions. Sir Humphrey Davy was requested to find a solution to this problem, and in 1815 he presented a paper to the Royal Society of London entitled "On the Fire-Damp of Coal Mines, and on Methods of Lighting a Mine so as to Prevent its Explosion." This resulted in the invention of the famous Davy lamp that uses a fine metal gauze as a flame arrester. He demonstrated that a metal gauze having about 28 openings per linear inch would cool the products of combustion so that a flame would not ignite flammable gas on the other side of the gauze. To avoid danger resulting from failure of a single gauze cylinder surrounding the flame, he found it necessary to use two concentric cylinders, one slightly smaller than the other. The lower edges were fitted snugly to the bowl containing the fuel, and the upper ends of the cylinders were covered by disks of similar gauze.

Also in 1815, but before Davy presented his first lamp to the public, George Stephenson (one of the pioneers in the development of the steam locomotive) quite independently was also working on a safe miner's lamp. He discovered during his experiments that flame produced by a particular gas at a given concentration will not pass through a tube smaller than a certain diameter. While most people have heard of Davy's lamp (it seems that Sir Humphrey received all the credit), it was actually Stephenson's discovery that was extremely important, because it provided the basis for the concept of the "quenching distance," which in turn led to the concept of "Maximum Experimental Safe Gap" (MESG). The MESG is extensively used today to classify gases and gas mixtures for the purpose of selecting flame arresters and electrical equipment. For further discussion of historical developments, see Smiles (1975).

Flame arresters for chemical process equipment and flammable liquid containers have been available for over 120 years. A US patent was issued as early as 1878 for a "spark-arrester" (Allonas 1878), while another "spark-arrester" was patented in 1880 (Stewart 1880). Numerous US patents have been issued for various designs of flame arresters, with one as recent as 1995 (Roussakis and Brooker 1995). In Germany, patents were issued in 1929 and 1939 for flame arresters that contained shock absorber internals upstream of the flame arrester elements. This innovation made them suitable as detonation arresters (Wauben 1999).

The crimped metal ribbon flame arrester element (see Chapter 5), which is used in both deflagration and detonation flame arresters, was the concept of Mr. Swan, an RAF Engineering Officer, who worked at the Royal Aircraft Establishment at Farnborough, England (Binks 1999). He needed a flame arrester for use during purging of gas bags of dirigibles, which then used flammable hydrogen rather than the inert helium used today. In this application it was used as a deflagration flame arrester, although it was also used as a detonation flame arrester for Group D and C gases (Group IIA and IIB in Europe). These classifications will be discussed in Section 5.3.1. Mr. Swan's crimped metal ribbon arresters were licensed to IMI Amal in the UK and were first manufactured in the late 1910s or early 1920s. There were many applications for crimped metal ribbon flame arresters during World War II in aircraft and motor torpedo boat engines, but they were mostly used as deflagration flame arresters. Their widespread use in detonation flame arresters occurred after World War II.

It was not until the 1950s that detonation flame arresters made of crimped metal ribbon elements were developed and began to be used more frequently (Binks 1999). The major impetus for the use of crimped metal ribbon detonation flame arresters in the US was the enactment of clean air legislation (Clean Air Act of 1990) which inadvertently created a safety problem by requiring reductions in volatile organic compound (VOC) emissions. To do this, manifolded vent systems (vapor collection systems) were increasingly installed in many chemical process industry plants which captured VOC vapors and transported them to suitable recovery, recycle, or destruction systems. This emission control requirement has led to the introduction of ignition risks, for example, from a flare or via spontaneous combustion of an activated carbon adsorber bed. Multiple connections to a flare header greatly increase the variability of the mixture composition and the probability of entering the flammable range. In addition, the long piping runs in manifolded vent systems contribute to the greater potential for deflagration-to-detonation transitions. Therefore, awareness of the need for detonation flame arresters for in-line applications did not gain widespread acceptance until the early 1990s.

2.2. Case Histories of Successful and Unsuccessful Applications of Flame Arresters

2.2.1 Successful Applications

Successful case histories of flame arrester applications are not commonly reported since no damage to equipment or loss of life occurred. However, one case history was found in the technical literature, which is presented below.

Sutherland and Wegert (1973) reported details of an acetylene decomposition incident in which a hydraulic flame arrester was used. The incident involved two plants that were connected by an 8-inch pipeline transporting acetylene at a pressure of about 14 psig. The acetylene distribution network consisted of approximately 2500 feet of 6-inch pipe in one plant; 5.5 miles of 8-inch pipe running between the two plants; and about 6000 feet of 8-inch, 6-inch, 5-inch, 4-inch, and 3-inch pipe within the other plant. A fire developed in the plant receiving the acetylene and resulted in an acetylene decomposition within the acetylene distribution network, which rapidly developed into a detonation. This progressed throughout the 7 miles of acetylene distribution piping in approximately 6 seconds. Fortunately, hydraulic flame arresters had been installed in the acetylene producing unit distribution line, in all user branch lines in both plants, and in the line to the flare stack in the plant supplying the acetylene. Except for disruption of flows, little mechanical damage was done to either plant as the hydraulic flame arresters functioned as designed. Also, the design of the acetylene piping to withstand detonations contributed to the mitigation of damage. However, because acetylene decomposes, forming hydrogen plus soot, the aftermath of the incident involved substantial cleaning and recommissioning efforts.

Kirby (1999) reports two successful applications of deflagration flame arresters. In one incident, a deflagration flame arrester was installed near the junction of a collection header from an ethylene oxide process unit with a flare stack. Although this type of flame arrester was really inappropriate for this service, the flame arrester satisfactorily interrupted a flashback from the flare stack into the header. A loud noise was heard, but an explosion did not propagate upstream of the arrester. The mixture in the header was barely within the flammable limits. The second incident involved a deflagration flame arrester installed near an incinerator, which satisfactorily stopped flame propagation back into a header containing acrylonitrile and air–nitrogen atmosphere. A standing flame occurred at the outlet of the flame arrester, and a thermocouple automatically actuated a valve which interrupted flow so that the flame went out.

2.2.2. Unsuccessful Applications

Case histories of unsuccessful applications of flame arresters are found more often in the open technical literature. Ten flame arrester (parallel plate and crimped metal ribbon) failure incidents in the Canadian petroleum industry are presented and their causes examined by Roussakis and Lapp (1989). According to the authors these incidents, which occurred between 1979 and 1989, represent only a small fraction of explosions that have taken place every year in the province of Alberta and were not prevented because the flame arresters were faulty. All of the ten flame arrester failures discussed occurred in sour oil vapor flaring operations. Nine of them involved oil production sites, and one involved a disposal facility for salt water contaminated during sour oil production. Eight of the ten flame arrester failures resulted in explosions and fires in oil storage tanks. One failure, involving flashback into a knockout tank, sent the gauge float soaring into the air as a high velocity projectile. Another incident resulted in the explosive destruction of a pair of concrete saltwater holding tanks. Eight of the incidents took place during startup, one took place during emergency shutdown caused by a power outage, and the other occurred during shut down for repairs. The reasons for the failures were categorized as follows:

1. Some arresters were used under conditions that exceeded their test limitations.
2. Some were not subjected to any official testing.
3. Some units failed because they had working channels that were distorted by the flame front overpressure.
4. Some failed due to structural or design flaws that provided a flame pathway through the element.

Howard (1988) reported a flame arrester failure in a major chemical plant with a large partial oxidation unit. The desired product was recovered in an absorber, and the off gas containing traces of organics was sent to an incinerator. To prevent flashback of flame from the incinerator into the upstream equipment, the engineering contractor designed and installed a horizontal flame arrester with conical inlet and outlet sections and a flame arrester section filled with packing. For the 4-foot-diameter pipeline to the incinerator, the flame arrester was a 12-foot-diameter cylinder with conical inlet and outlet sections, and the arrester element was a 3-foot-wide section filled with 2-inch Pall rings. The total included angle of the inlet and outlet cones was about 150°. The design was flawed since it was known from experiments that 2-inch Pall rings will not stop a flame from combustion of hydrocarbon–air mixtures that are well within the flammable range. Also, the wide cone angles of the arrester are not conducive to good flow distribution over the entire cross section of the Pall ring packing. When a process upset caused flashback from the incinerator, the flame arrester failed to stop the flame, which resulted in damage to the equipment on the other side of the arrester.

A chemical plant flare experienced a series of three consecutive deflagrations resulting in severe damage to the flare water seal flame arrester (Desai 1996). The deflagrations occurred during process startup after a complete unit shutdown. The first flashback from the flare is believed to have tilted the water seal internals such that it lost its effectiveness as a flame arrester. After operators reset both the natural gas and snuffing steam interlocks, the methane flow was reestablished to the flare, and steam flow was stopped. This resulted in a second flashback that was felt as a "rumble" by the operators. However, at this point, the water seal flame arrester was believed to be nonfunctional (as evidenced by recorded low pressure drop), allowing the flashback to travel into the knockout drum and cause an explosion. Production personnel once again reset the interlocks, and the same flashback scenario was repeated for a third time. This deflagration probably caused additional internal damage to the water seal. Overall damage was limited to the water seal only, and no one was injured.

Fabiano et al. (1999) describe an explosion in the loading section of an Italian acetylene production plant in which the installed flame arresters did not stop a detonation. The arresters were deflagration type and the arrester elements were vessels packed with silica gel and aluminum plates (Fabiano 1999). It was concluded that the flame arresters used were not suitable for dealing safely with the excess pressures resulting from an acetylene decomposition, and may not have been in the proper location to stop the detonation.

2.3. Evolution of Standards and Codes

This section is an overview of the evolution of standards and codes for flame arresters from various organizations in the United States, Canada, the United Kingdom, and European countries. Brief summaries of the contents and requirements of various standards and codes are presented in Chapter 8.

2.3.1. United States

Underwriters Laboratories (UL) issued the first edition of their standard for flame arresters, UL 525, in December 1936. Subsequent revisions were published in July 1946, September 1973, June 1979, September 1984, and the latest revision (6th edition) in December 1994. It was also issued as an ANSI standard in 1995. The first five editions applied only to end-of-line deflagration arresters mounted on atmospheric pressure storage tanks. The sixth edition represents a major revision of the previous requirements for deflagration flame arresters (Part II) and also contains a new section (Part III) on detonation flame arresters. According to the scope of UL 525 the requirements for tank vent deflagration flame arresters cover arresters for use on vents of storage tanks for petroleum oil and gasoline only. The requirements for in-line detonation flame arresters cover arresters for use in piping systems containing flammable vapors and gases in mixture with air.

The United States Coast Guard (USCG) on June 21, 1990, published requirements for facilities that collect vapors of flammable or combustible chemicals emitted from barge and ship cargo tanks (Title 33 CFR, Part 154, Subpart E). Amendments were issued in 1996 and 1998. Specific requirements are given only for the collection of vapors of crude oil, gasoline, and benzene. However, the regulations can be, and are being applied for other chemicals if the facility is approved by the Commandant of the Coast Guard as meeting the requirements. The USCG has internal guidelines that are used by their personnel to evaluate whether a facility meets the requirements for any other chemicals (Schneider 1999). Title 33 CFR, Part 154 is being revised at the present time to cover many other chemicals (Schneider 2000). Appendix A to Part 154 presents guidelines for detonation flame arresters, and Appendix B to Part 154 is a standard specification for tank vent flame arresters.

Factory Mutual Research (formerly Factory Mutual Research Corporation) published a test procedure for flame arresters for storage tank vent pipes in 1990 (FMRC 1990). The Factory Mutual Research test protocol adopted the USCG standard, but modified it with respect to the endurance burn test. Factory Mutual Research also published in 1999 an examination program for testing of detonation arresters, which is also based on the USCG requirements (FMRC 1999).

The American Society for Testing and Materials (ASTM) first published a specification for tank vent flame arresters, F1273, in 1991 (ASTM 1991). This was reaffirmed in 1996. This specification provides the minimum requirements for design, construction, performance, and testing of tank vent flame arresters (deflagration flame arresters).

The API has published two recommended practices on flame arresters. The first one, RP 2210, was issued as a second edition in 1982, and a revision (3rd edition) was published in 2000. It covers flame arresters for vents of tanks storing petroleum products, including sections on background, flame arrester construction, application, problems, maintenance, and listing; also, pressure-vacuum valves as a substitute for flame arresters. The second one, Publication 2028, was originally published in 1984, issued as a second edition in 1991, and a third edition is expected to be published in 2002. It is concerned with flame arresters in piping systems and discusses flame propagation, listed flame arresters (topics include problems in piping systems, design, limitations, test procedures, untested arresters, and flame arresters in series), and unlisted arresters (topics include water seals, packed beds, velocity-type arresting devices, and mechanical interruption of flame path). Both of these publications contain useful information, but are rather limited in their coverage of the various topics included.

2.3.2. Canada

The Canadian Standards Association published a standard on test methods for in-line and firebox flame arresters in 1991. A revision was published in 1998 (CSA 1998).

2.3.3. United Kingdom

A manual for testing deflagration flame arresters was published by the Fire Research Station in 1978 (Rogowski 1978). In 1990 the British Standards Institute published BS 7244, which specifies requirements for deflagration and detonation flame arresters suitable for general use in venting vessels, storage or transport tanks, protection of the fuel/air supply to gas burners, and in industrial plants which handle gases or vapors with autoignition temperatures below 200°C (392°F), using the specified UK test method. It describes test methods that are used to classify arresters in terms of their performance so that they can be correctly matched to the intended application (BSI 1990). Thomas and Oakley (1993) point out some practical difficulties encountered when testing deflagration and detonation flame arresters according to BS 7244. A new standard, BS EN 12874:2001, which is actually the CEN standard, has been in issued in 2001.

2.3.4. Europe and International

In Germany, deflagration and detonation flame arresters are tested and approved by the German Federal Physical-Technical Institute (PTB) in Braunschweig and the German Federal Institute for Material Research and Testing (BAM) in Berlin. However, once the new European Committee for Standardization (CEN) standard is approved, it will supersede the PTB and BAM procedures.

A new CEN European standard, EN 12874, was issued in 2001. It will be used by all European nations that belong to the European Union (EU), as well as other non-EU nations that are members of the CEN. This is a very comprehensive standard covering many aspects of flame arrester technology, including specific requirements for the following:

- static element flame arresters (fixed element standard dry type)
- liquid product detonation flame arresters (in liquid filled or partially filled liquid piping)
- high velocity vent valves
- flow controlled apertures (velocity flame stoppers)
- hydraulic flame arresters (liquid seals for gas/vapor lines)
- testing of flame arresters in equipment (rotary pumps, compressors, blowers, fans, and vacuum pumps)
- flame arresters combined with breather valves (conservation vent valves)

Flame transmission test procedures for deflagration, detonation, and burning tests are discussed in all of the above sections.

The International Maritime Organization (IMO) published a standard in 1984 for the design, testing and locating of deflagration and detonation flame arresters and high velocity vent valves for cargo tanks in tanker ships (IMO 1984). This was amended in 1988 and reissued as Revision 1 (IMO 1988). A new revised standard was issued in 1994 (IMO 1994).

2.4. Safety Concerns and Environmental Regulations: Tradeoffs and Conflicts

The primary concern about hazardous processes is that they be operated in a safe manner so that fires, deflagrations, detonations, and releases of flammable and toxic gases into the atmosphere or inside process buildings are eliminated or minimized. Also of great concern from an environmental standpoint is the control and minimization of volatile organic compound (VOC) emissions from process vents and fugitive sources.

The US Clean Air Act of 1970 and its Amendments of 1977 and 1990 require the control of emissions of hazardous VOCs in most chemical processes. This has resulted in the manifolding of vents from storage tanks and process equipment containing hazardous chemicals, and the routing of the manifolds to control equipment such as flares, thermal and catalytic oxidizers, activated carbon adsorbers, and scrubbers. Although this has resulted in an improvement in environmental conditions, it has created an increased safety problem (CCPS 1993, Ozog and Erny 1999). Vent manifolding has now made it possible for a flame to propagate from one source to multiple items of equipment connected to the vent manifold. If the vent manifold piping is sufficiently long, a deflagration can develop into a detonation. Therefore, consideration should be given to installing detonation flame arresters at proper locations in the manifold system, or to providing other protective measures (see Chapter 3). If detonation flame arresters are used, they must be tested according to an approved test protocol. Also, manifolded vent systems should be analyzed using an appropriate hazard analysis method.

2.5. References

2.5.1. Regulations, Codes of Practice, and Industry Standards

The editions that were in effect when this book was written are indicated below. Because standards and codes are subject to revision, users are encouraged to apply only the most recent editions.

API PB 2028. 2002. *Flame Arresters in Piping Systems,* 2nd ed. American Petroleum Institute, Washington, D.C.

API PB 2210. 2000. *Flame Arresters for Vents of Tanks Storing Petroleum Products,* 2nd ed. American Petroleum Institute, Washington, D.C.

ASTM F-1273-91. 1991. *Standard Specification for Tank Vent Flame Arresters.* American Society for Testing and Materials, West Conshohocken, PA.

BSI (British Standards Institution) BS 7244. 1990. *Flame Arresters for General Use.* British Standards Institution, London, England, UK.

CEN (Comité Européean de Normalisation) EN 12874. 2001. *Flame Arresters—Specifications, Operational Requirements and Test Procedures.* European Committee for Standardization, Brussels, Belgium.

CSA (Canadian Standards Association) CSA-Z343. 1998. *Test Methods for In-Line and Firebox Flame Arresters.* Canadian Standards Association, Toronto, Ontario, Canada.

FMRC (Factory Mutual Research Corporation) Class 6061. 1990. *FMRC Test Procedure for Flame Arresters for Storage Tank Vent Pipes.* Factory Mutual Research Corporation, Norwood, MA.

FMRC (Factory Mutual Research Corporation) Class 6061. 1999. *Examination Program, Detonation Flame Arresters.* Factory Mutual Research Corporation, Norwood, MA.

IMO (International Maritime Organization) MSC/Circ. 373. 1984. *Standards for the Design, Testing and Locating of Devices to Prevent the Passage of Flame into Cargo Tanks in Oil Tankers.* International Maritime Organization, London, England, UK.

IMO (International Maritime Organization) MSC/Circ. 373/Rev. 1. *Revised Standards for the Design, Testing and Locating of Devices to Prevent the Passage of Flame into Cargo Tanks in Tankers.* International Maritime Organization, London, England, UK.

IMO (International Maritime Organization) MSC/Circ. 677. 1994. *Revised Standards for the Design, Testing and Locating of Devices to Prevent the Passage of Flame into Cargo Tanks in Tankers.* International Maritime Organization, London, England, UK.

33 CFR Part 154, Subpart E-Vapor Control Systems. 1990. U.S. Department of Transportation, U.S. Coast Guard.

UL 525. 1994. *Standard for Flame Arresters.* 6th edition. Underwriters Laboratories, Inc., Northbrook, IL. (Also available as ANSI/UL 525-1995.)

2.5.2. Specific References

Allonas, J. 1878. U.S. Patent No. 295,716. *Spark-Arrester* (July 9, 1878).

Binks, G. 1999. In-house publication of IMI Amal. Personal communication to S. S. Grossel by D. Long, IMI Amal (June 28, 1999).

CCPS. 1993. *Guidelines for Engineering Design for Process Safety.* American Institute of Chemical Engineers, Center for Chemical Process Safety, New York, NY.

Desai, V. M. 1996. A Flare Deflagration Incident at Rohm and Haas. *Process Safety Progress,* 15(3), 166–167.

Fabiano, B. 1999. Personal communication from B. Fabiano, Department of Chemical and Process Engineering, University of Genoa, Italy to S. S. Grossel (November 26, 1999).

Fabiano, B., Pastorino, R., and Solisio, C. 1999. Explosion at an Acetylene Plant: A Methodological Approach to Accident Analysis. *IChemE Loss Prevention Bulletin,* Issue 145 (February 1999).

Howard, W. B. 1988. Process Safety Technology and the Responsibility of Industry. *Chemical Engineering Progress,* 84(9), 25–33 (September 1988).

Kirby, D. C. 1999. Personal communication from D. C. Kirby, Union Carbide Corporation to S. S. Grossel (December 8, 1999).

Ozog, H., and Erny, W. J. 1999. Safety Hazards Associated with Air-Emission Controls. Paper 5d. *Proc. 33rd Annual AIChE Loss Prevention Symposium,* March 14–17, 1999, Houston, TX. American Institute of Chemical Engineers, New York, NY.

Rogowski, Z. W. 1978. *Manual for Testing Flame Arresters.* Fire Research Station, Building Research Establishment, Borehamwood, Herts., England, UK.

Rousssakis, N. and Brooker, D. E. 1995. U.S. Patent No. 5,415,233, *Arrestor Apparatus* (May 16, 1995).

Roussakis, N. and Lapp, K. 1989. A Summary of Investigations from Ten Inline Flame Arrester Failures. Paper presented at a meeting of the API Committee on Safety and Fire Protection, September 21, 1989, Seattle, WA.

Schneider, A. 1999. Personal communication from Dr. Alan Schneider, U.S. Coast Guard, to S. S. Grossel (June 2, 1999).

Schneider, A. 2000. Personal communication from Dr. Alan Schneider, U.S. Coast Guard, to S. S. Grossel (April 12, 2000).

Smiles, S. 1975. *The Lives of George and Robert Stephenson.* The Folio Society, London, England, UK.

Stewart, G. 1880. U.S. Patent No. 223,403, Spark-Arrester (January 6, 1880).

Sutherland, M. E., and Wegert, H. W. 1973. An Acetylene Decomposition Incident. *Chemical Engineering Progress,* 69(4), 48–51.

Thomas, G. O., and Oakley, G. L. 1993. On Practical Difficulties Encountered When Testing Flame and Detonation Arresters to BS 7244. *Trans. IChemE,* 71, Part B, 187–193.

Wauben, M. 1999. Personal communication from M. Wauben, SchuF (USA), Inc. to S. S. Grossel (June 17, 1999).

3

Overview of Deflagration and Detonation Prevention and Protection Practices

3.1. Introduction

This chapter presents an overview of the various practices used in the chemical process industries to prevent and protect against deflagrations and detonations. It includes an introduction to deflagration and detonation flame arresters and also other methods that can be used when flame arresters are not practical or are too expensive.

3.2. Deflagration and Detonation Flame Arresters

Flame arresters are broadly divided into two major types: deflagration and detonation flame arresters. Another parameter for selecting a flame arrester is directionality, which refers to the direction of flame approach for which an arrester is designed to operate in a pipeline. The pipeline connecting a flame arrester with an identified ignition source is the "unprotected side" of the arrester. The pipeline connecting the arrester with at-risk equipment (equipment that must be protected from the temperature or pressure associated with flame penetration) is the "protected side." If a flame arrester may encounter a flame arriving from only one direction, a "unidirectional" arrester can be used. If a flame may arrive from either direction, a "bidirectional" flame arrester is needed. The latter arrester is either symmetrically constructed or has been tested and certified for deflagrations and detonations approaching from either direction. Back-to-back use of unidirectional flame arresters will not usually be cost-effective unless testing reveals a specific advantage such as increased allowable operating pressure during restricted-end deflagration testing where the test pipeline has one end closed.

A number of essential points about flame arresters are as follows (CCPS 1993):

1. A flame arrester is a device permeable to gas flow but impermeable to any flame it may encounter under anticipated service conditions. It must both quench the flame and cool products sufficiently to prevent reignition at the arrester outlet.
2. Proper application of a flame arrester can help avoid catastrophic fire and explosion losses by providing a flame barrier between at-risk equipment and anticipated ignition sources.
3. Flame arresters have often failed in practice. There have been significant advances in flame arrester technology over the last decade that explain many "failures" as due to misapplication. Plant inspections have shown that misapplication of flame arresters continues to be common.
4. Flame arresters cannot be designed from first principles and can only be proven by tests simulating the conditions of use. The user should ensure that a flame arrester has been properly tested to meet the intended purpose, and should be prepared to stipulate the required performance standard or test protocol to be followed.
5. In almost all cases, if a flame arrester is placed in-line rather than at (or close to) the open end of a vent pipe, a detonation flame arrester is needed. Detonation flame arresters must be able to stop both deflagrations and detonations. They require extensive testing and mandatory testing protocols may apply.
6. Unless flame arresters meet all the necessary requirements discussed in this book, they should be replaced, relocated, or eliminated in favor of an alternative means of protection.

Flame arresters are classified according to certain characteristics and operational principles, as follows:

- location in process
- combustion conditions (deflagration or detonation, operating temperature and pressure)
- arrester element (matrix) construction for dry type arresters

- other types (hydraulic liquid seal, high velocity vent valves and velocity flame stoppers)

These are briefly discussed below.

Location in Process

Flame arresters are classified according to their location with respect to the equipment they are protecting. When a flame arrester is located directly on a vessel/tank vent nozzle, or on the end of a vent line from the vent nozzle, it is called an end-of-line arrester, and is usually a deflagration flame arrester. These are commonly installed on atmospheric pressure storage tanks, process vessels, and transport containers. If the vented vapors are ignited, perhaps by lightning, the flame arrester will prevent the flame from spreading from the atmosphere to the inside of the vessel. Such flame arresters are neither designed for nor suitable for stopping detonations. Figure 3-1 shows the installation of end-of-line flame arresters on vessels located inside of buildings and outdoors.

If the arrester is not placed at the end of a line, it is known as an in-line arrester. In-line arresters can be of the deflagration or detonation type,

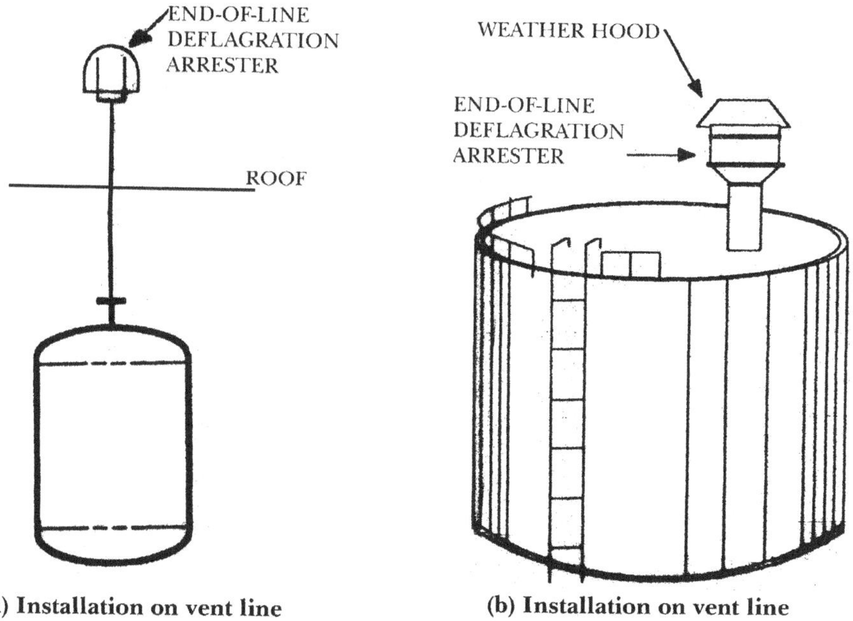

(a) Installation on vent line from indoor vessel

(b) Installation on vent line from outdoor tank

FIGURE 3-1. End-of-line deflagration flame arrester locations.

depending on the length of piping and pipe configuration on the unprotected side of the arrester and the restrictions on the protected side of the arrester. A detonation flame arrester is used in all cases where sufficient "run-up" distance exists for a detonation to develop. Some types of in-line deflagration flame arresters are called "pipe-away" or vent-line arresters. These are also installed on the vent nozzles of atmospheric pressure tanks and vessels, and have a short length of pipe attached to them to direct the vapors, and possibly flame, away from the tank or vessel roof. Figure 3-2 shows the installation of an in-line deflagration flame arrester. The maxi-

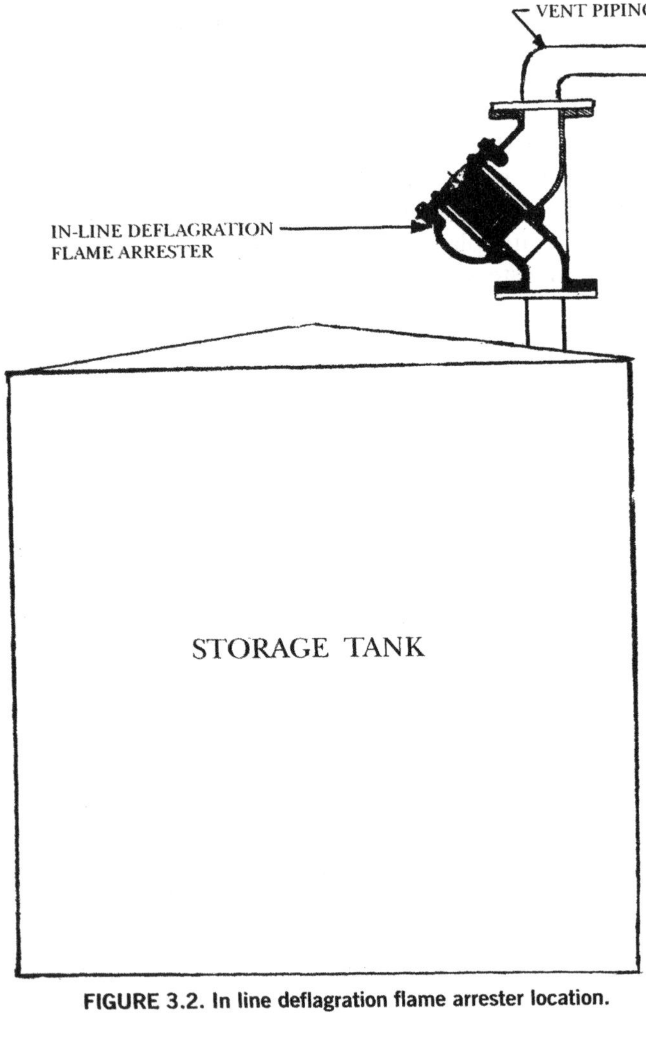

FIGURE 3.2. In line deflagration flame arrester location.

mum length of pipe from the discharge side of an in-line flame arrester to the atmosphere is usually no more than 20 feet for Group D gases, and is a function of the pipe size and the manufacturer's design. For other gases (Group B or Group C), the maximum allowed distance must be established by proper testing with the appropriate gas mixture and the pipe diameter involved. Turbulence-promoting devices in the flow path (bends, elbows, tees, valves, etc.) cannot be used unless testing has addressed the exact geometry in the installation. It is essential to ensure that run-up to detonation cannot occur in the actual piping system. For some fast-burning gases such as hydrogen–air and ethylene oxide–air, the run-up distance (maximum attached pipe length) can be appreciably less. In all cases the manufacturer should be consulted for their recommended maximum "run-up" length for in-line deflagration flame arresters.

An in-line detonation flame arrester must be used whenever there is a possibility of a detonation occurring. This is always a strong possibility in vent manifold (vapor collection) systems, where long pipe runs provide sufficient "run-up" distances for a deflagration-to-detonation transition to occur. Figure 3-3 shows the installation of in-line arresters of the detonation type in a vent manifold system.

Combustion Conditions

As mentioned earlier, flame arresters can be classified as either deflagration or detonation type. Deflagration flame arresters on tanks are designed to stop a flame from propagating into a tank from an unconfined atmo-

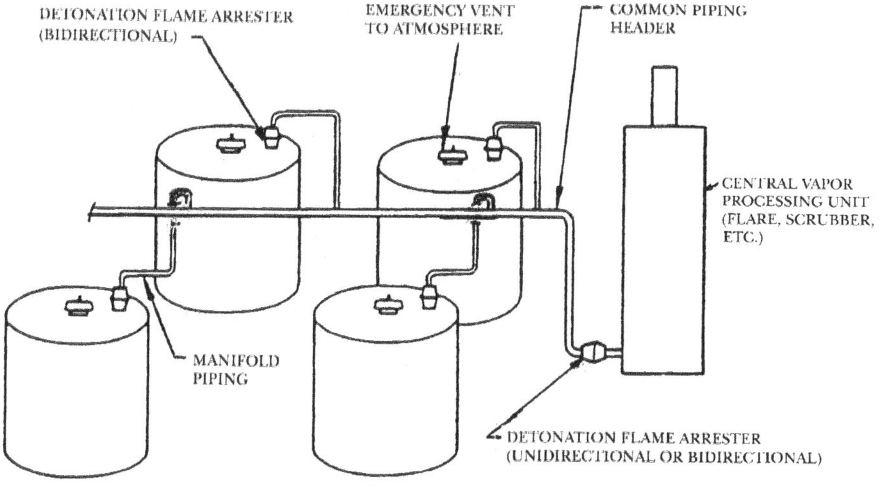

FIGURE 3.3. In-line detonation flame arrester locations.

spheric deflagration, or to prevent a flame generated from a confined volume deflagration in a vessel from escaping to the outside of the vessel (Halstrick 1995). They normally cannot withstand significant internal pressure and cannot stop detonations. Typical flame speeds in a deflagration occurring in piping range from 10 to 200 ft/s. Deflagrations of fuel–air mixtures typically generate pressures 8 to 12 times the initial pressure in closed process vessels and equipment. Figure 3-4 shows several types of deflagration arresters.

Detonation flame arresters are devices designed to withstand and extinguish the high speed and high pressure flame front that characterizes a detonation propagating through a piping system. Therefore, a detonation arrester must be able to withstand the mechanical effects of a detonation shock wave while quenching the flame. Some designs have a "shock absorber" in front of the flame arresting element to reduce both the high pressure shock wave and the dynamic energy and to split the flame front before it reaches the flame arrester element (Halstrick 1995). Figure 3-5 shows details of a "shock absorber" as designed by Protego™. Another design variation has what is called a "detonation momentum attenuator" (DMA) (Westech 1989). Details of such a DMA as designed by Westech is shown in Figure 3-6. Detonations occurring in piping have velocities of about 6000 ft/s, or greater, and in closed process vessels and equipment can generate pressures from 20 to 100+ times the initial pressure. Detonation flame arresters are available as both unidirectional or bidirectional types. When installed in a vent manifold system the flame arresters on the tanks may be unidirectional or bidirectional, depending on the manufacturer's recommendations. They should preferably be installed in a vertical orientation, so that if liquid is present, the arrester will drain. If they must be installed in a horizontal orientation, they should be provided with drain connections. Most detonation arresters have crimped metal ribbon arrester elements, although expanded metal cartridges are also used. Arrester elements for detonation arresters are usually longer than for deflagration arresters. Figure 3-7 shows several designs of detonation arresters.

It has been reported that there have been some cases where a detonation arrester failed to stop a deflagration (Howard 1982). This occurs when a restriction (e.g., a valve) exists on the protected side of the arrester. Roussakis and Lapp (1991) present data from tests on three types of in-line flame arresters that corroborates this seemingly anomalous behavior. The causes for this phenomenon are complex and depend on such factors as run-up length effect on relative overpressure (ratio of pressure rise caused by the flame front to the absolute operating pressure at the time of ignition), flow restriction on the protected side, and the absolute operating pressure. Flame quenching capability of a flame arrester is determined by

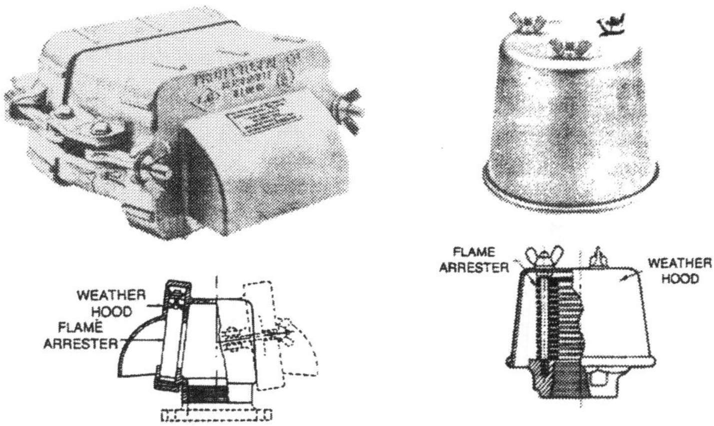

(a) End-of-line parallel plate flame arresters

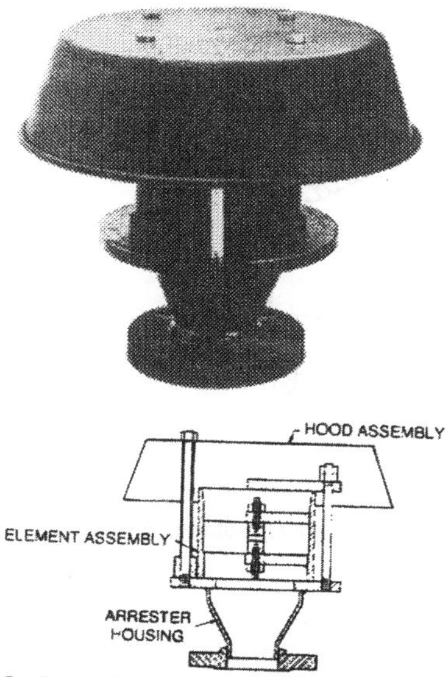

(b) End-of-line crimped metal ribbon flame arrester

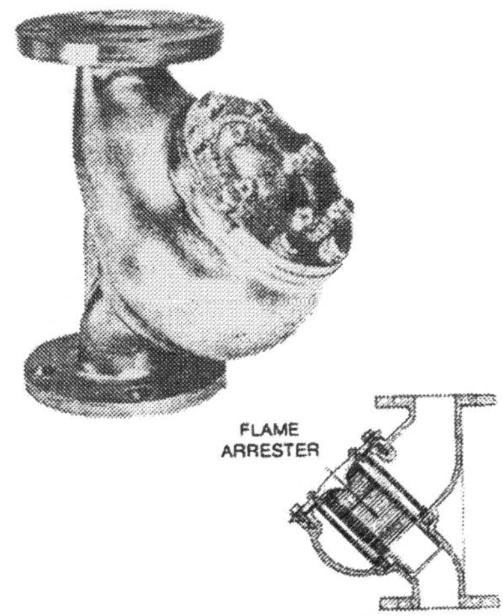

(c) In-line parallel plate flame arrester

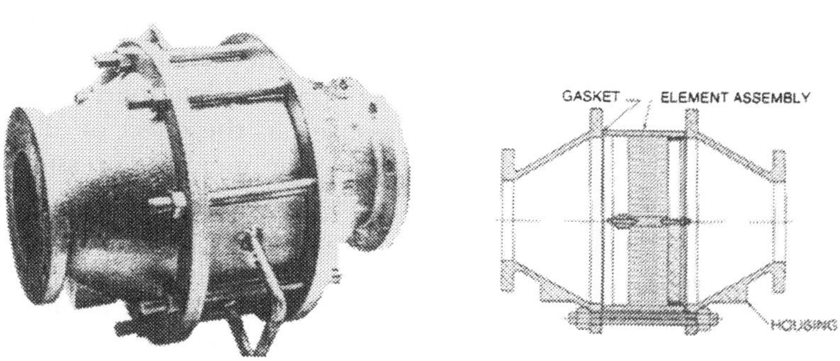

(d) In-line crimped metal flame arrester

FIGURE 3-4. Types of deflagration flame arresters.

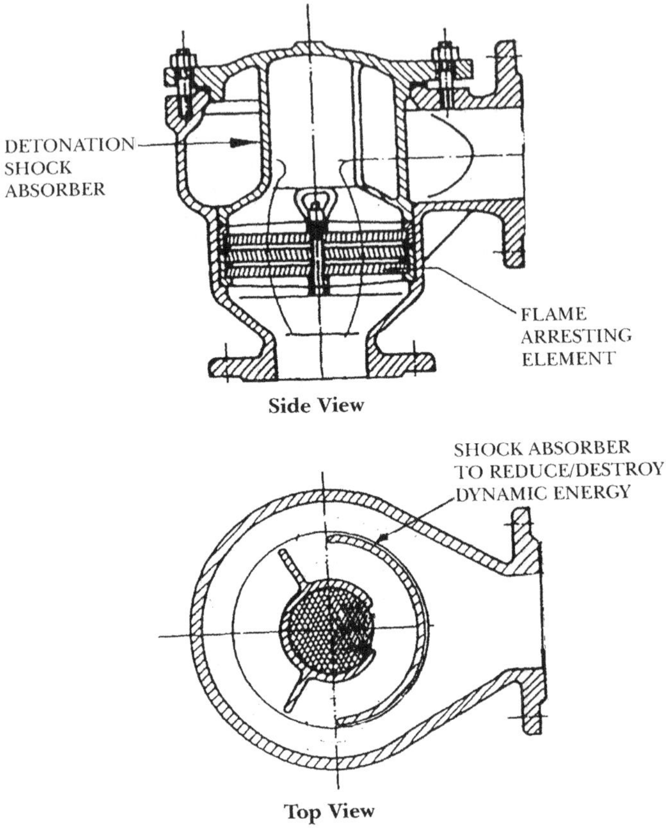

FIGURE 3-5. Protego detonation flame arrester with shock absorber. (Source: Protego/Braunschweiger Flammenfilter GmbH)

the initial operating pressure. The following explanation is given in Canadian standard CSA-Z343 (1998). In the low and medium pressure deflagration zone, a "reflected pressure" effect may take place, which reduces the maximum operating pressure in the flame arrester (the pressure at which the flame arrester would have successfully quenched a flame if there was no reflected pressure effect) when ignitions occur in that zone. In essence, what happens is that the reflected pressure has the effect of allowing a flame to pass through a flame arrester at a lower operating pressure. Consider the case of a given hypothetical flame arrester that is able to successfully stop detonations when the initial pressure (the pressure in the pipe just before the explosion) is less than 20.3 psia. If a low or medium pressure deflagration was involved rather than a detonation, the flame arrester could fail at an initial pressure of less than 20.3 psia.

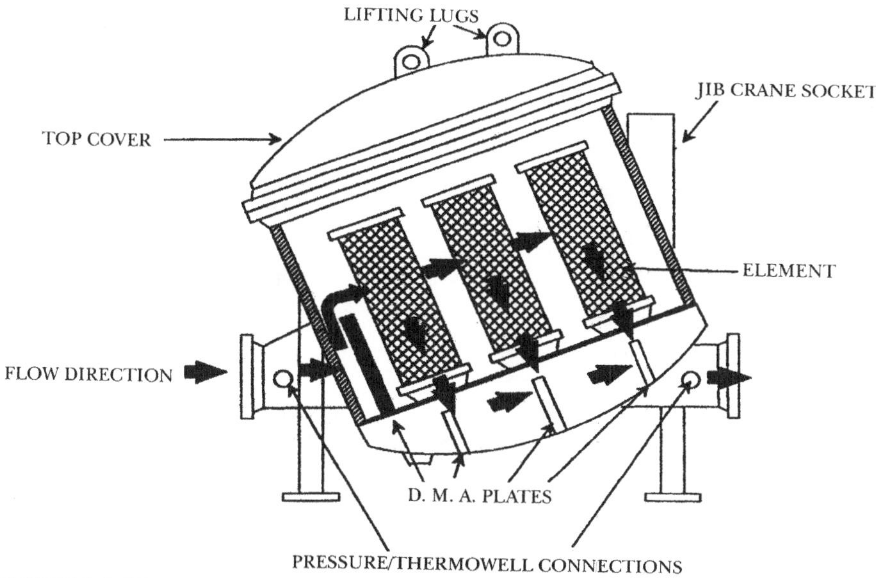

FIGURE 3-6. Westech detonation flame arrester with detonation momentum attenuator (D.M.A). (Source: Westech Industrial, Ltd.)

During a deflagration, unburned gas ahead of the flame is precompressed, leading to an increase of pressure in the arrester before the flame arrives. This increased pressure may facilitate passage of the flame through the arrester element if it significantly exceeds the pressures that were developed during certification testing. The buildup of pressure in the flame arrester is influenced by downstream obstructions that are addressed by restricted-end deflagration testing as described in Chapter 8. The following is one possible scenario for reflected pressure effects. When the gas is ignited, two flame fronts propagate from the ignition point—one front moves down the pipe toward the flame arrester, and one moves up the pipe away from the flame arrester. The flame front moving away from the flame arrester may have sufficient distance available along the pipe to accelerate and develop into a detonation. The detonation will produce the usual fast-rising pressure wave, which will eventually reach some obstruction as it continues to travel away from the flame arrester. When it does reach the obstruction, a wave of lesser pressure will be reflected back (as in an echo) toward the flame arrester. Traveling back toward the flame arrester, the reflected pressure wave will travel through burnt gas. During this period the other flame front, the one involving the relatively slow-moving deflagration, continues to progress toward the flame arrester. For

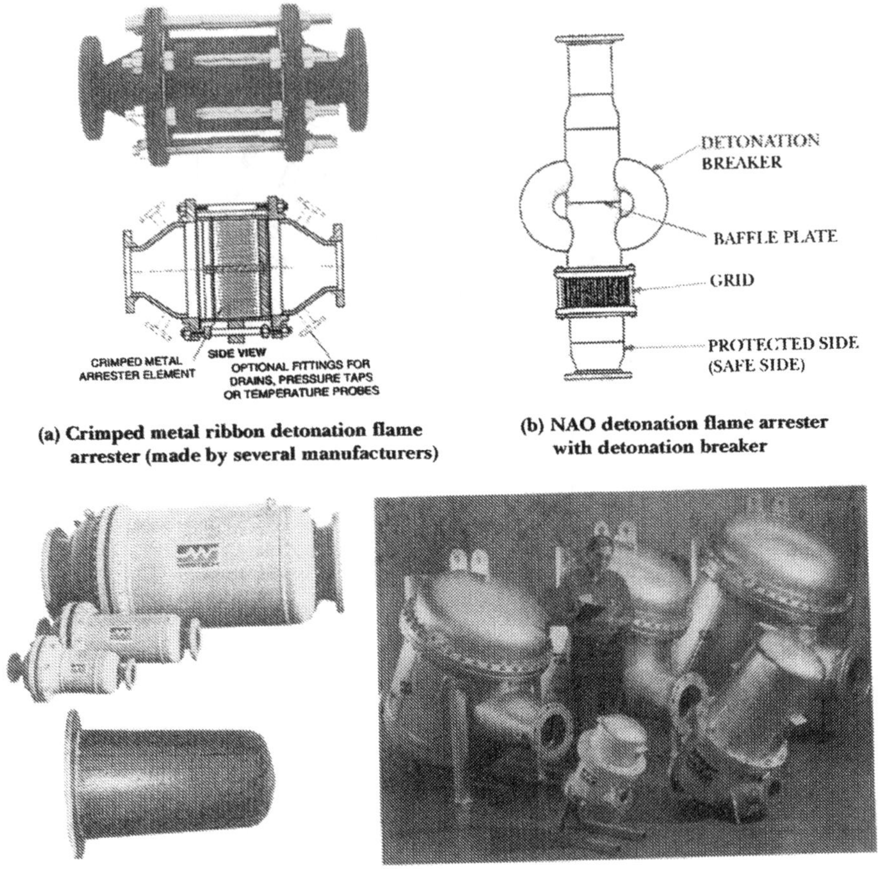

(a) Crimped metal ribbon detonation flame arrester (made by several manufacturers)

(b) NAO detonation flame arrester with detonation breaker

(c) Expanded metal cartridge detonation flame arrester (Westech Series 57)

(d) Expanded metal cartridge detonation flame arrester (Westech Series 55)

FIGURE 3-7. Various types of detonation flame arresters. (Sources: Protectoseal Company, NAO, Inc., and Westech Industrial, Ltd.)

certain adverse geometries, the coincident arrivals of the deflagration pressure wave and the reflected pressure wave at the flame arrester element in the unburned gas will have the effect of momentarily increasing the initial pressure at the element just prior to the arrival of the deflagration flame front. This transitory pressure increase may be sufficient to allow the flame front to pass through the flame arrester element. Another similar scenario involves the reflection coming from the protected side of the flame arrester. Because of this, before the flame arrester is specified and installed, realistic tests should be run using an approved test method (see Chapter 8) to verify that the flame arrester will stop both deflagrations and detonations.

Arrester Element (Matrix) Construction for Dry Type Arresters

Dry type deflagration and detonation flame arresters have an internal arrester element (sometimes called a matrix) that quenches the flame and cools the products of combustion. A great number of arrester elements have been developed and used. The most common types currently available are as follows:

- crimped metal ribbon
- parallel plate
- expanded metal cartridge
- perforated plate
- wire gauze and wire gauze in packs
- sintered metal
- metal shot in small housings
- ceramic balls

Other types that have been used but are not currently commercially available in the United States are compressed wire wool and metal foam. However, compressed wire wool flame arresters are available in the United Kingdom.

Currently available flame arrester elements will be discussed in detail in Chapter 5.

Other Types

Other types of deflagration and detonation flame arresters that do not contain an arrester element (matrix) have been used successfully in various applications where conventional dry type arresters were not suitable or were very expensive. Among these are:

- hydraulic (liquid seal)
- packed bed
- velocity flame stoppers
- high velocity vent valves
- conservation vent valves

These will be discussed in detail in Chapter 5.

3.3. Deflagration Venting

Venting of deflagrations is often used for overpressure protection for process equipment, pipes and ducts.

Procedures for the sizing and design of deflagration vents for pipes and ducts operating at or near atmospheric pressure (up to 3 psig) are presented in Chapter 8 of NFPA 68 (NFPA 1998).

The following guidelines are provided in Chapter 8 of NFPA 68:

1. Multiple vent locations can be provided along the length of a pipe or duct to reduce the maximum pressure generated during a deflagration.
2. Deflagration vents should be located close as possible to ignition sources where these sources can be identified.
3. Pipes or ducts connected to a vessel in which a deflagration can occur also need deflagration protection. Such protection can be accomplished by installing a vent on the pipe with an area equal to the cross-sectional area of the pipe or duct. It should be located on the pipe or duct no more than two pipe or duct diameters from the point of connection to the process vessel or equipment.
4. For systems that handle gases, vents should be provided on each side of turbulence-producing devices (e.g., elbows) at a distance of no more than three diameters of the pipe or duct.
5. In order to use the correlations presented later in the guide, the weight of the deflagration vent closures should not exceed 2.5 lb/ft^2 (12.2 kg/m^2) of free vent area.
6. The static burst pressure of the vent closures should be as far below P_{red} (the maximum pressure generated in the pipe or duct by the deflagration that one wishes to allow during venting) as practical and should be consistent with operating pressures.
7. Deflagration vents should discharge to a location that cannot endanger personnel.
8. Consideration should be given to reaction forces that develop during venting.

Guidelines are also given in NFPA 68 for vent placement to prevent deflagration-to-detonation transition. Methods are also presented for calculating the maximum pressure generated by a deflagration for a single deflagration vent and multiple deflagration vents on a pipe or duct, based on system flow velocities and gas fundamental burning velocities.

When deflagration venting is used, a major hazard of concern is the fireball (flame cloud consisting of burning gases and/or dust) discharged from the vent. This can cause harm to personnel or process equipment and

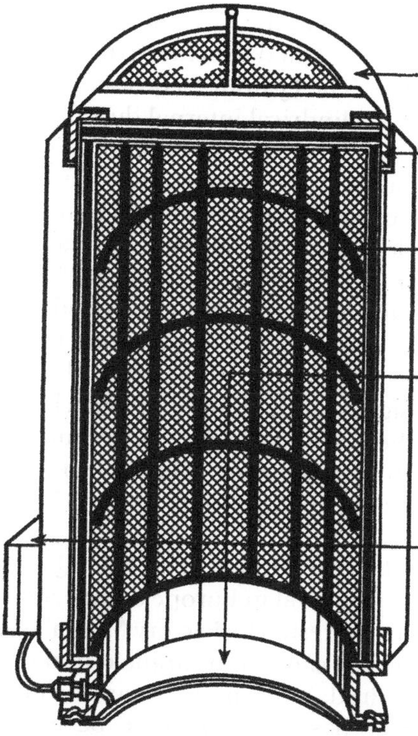

FIGURE 3-8. Combination flame-arresting and particulate retention vent system (Q-Rohr System). (Source: Cv Technology, Inc.)

buildings if it impacts on them. If the process equipment is inside of a building then a vent discharge duct is attached to the vent to direct the fireball outside of the building. An alternate solution to the fireball problem is the use of a combination flame-arresting and particulate retention vent system (see Figure 3-8, which shows details of such a device). As a deflagration is vented through the device, any burned and unburned dust is retained within the device, and the flame is quenched, so that no flame emerges from the equipment. In addition, near-field blast effects (overpressure) are greatly reduced outside the vented equipment. Its main advantage is that it can be installed on indoor equipment. This obviates the need for a vent duct that would otherwise be needed to direct the flame, combustion products, and particulate matter outside of the building.

Two types of such a device are currently commercially available, the Q-Rohr™ and the FlamQuench™. The Q-Rohr™ is manufactured by Rembe in Germany (available in the United States from Cv Technology, Inc.) and the FlamQuench II™ by Fike Corporation in the U.S. and Europe. The Q-Rohr™ and the FlamQuench II™ are designed to achieve the same thing, but have different mechanical designs. The Q-Rohr™ consists of a cylinder with a rupture disk at the inlet and a cylindrical internal dust filter (which retains the particles) of special ceramic-fiber mat and a second cylindrical internal flame arrester element of high-grade stainless steel mesh. On the other hand, the FlamQuench II™ consists of a cylinder with a rupture disk at the inlet and an internal cylinder consisting of particle-retaining and flame extinguishing stainless steel mesh layers. The Q-Rohr™ has been tested on a number of dust and dust-air-flammable gas hybrid mixtures (Stevenson 1998) and also for some specific gases (e.g., carbon monoxide, methane, propane, and alcohols), and these tests were all successful (Stevenson 2000). For gases with significantly different flame characteristics than the ones tested, further testing would be required. The FlamQuench II™ has been tested in vessels having volumes from 0.5 m^3 to 10 m^3 and with propane and dusts with K_{St} values up to 318 bar-m/s (Chatrathi 2001).

The Q-Rohr™ is approved by Factory Mutual. The FlamQuench II™ has been tested for CEN and ATEX approval in Europe at the Laboratorio Oficial Madariaga in Spain, and is being tested by Factory Mutual Global.

The deflagration venting approach can also be applied to the protection of vent manifold systems if desired. Venting products of combustion gases upstream of a flame arrester has been found to be effective in reducing the temperature measured on the run-up (unprotected) side of flame arresters for tests conducted with noncommercial flame arresters (White and Oswald 1992). Although venting appeared to have a small effect on peak overpressure and more effect on impulse, the benefit appears to be largely in a vent's ability to reduce the thermal impact on the flame arrester. Venting, however, presents environmental problems around the vent opening, which can include hot gas emissions, noise, and discharge of fragments (e.g., vent cover segments, hot particles). Properly located vents are designed to prevent a detonation from developing.

3.4. Oxidant Concentration Reduction

One of the most widely used methods of preventing deflagrations and detonations is oxidant concentration reduction. This method can be applied to process equipment and vent manifold systems. The prevention of deflagrations or detonations can be accomplished by either inerting or fuel enrichment.

In the case of inerting, the oxidant (usually oxygen) concentration is reduced by the addition of inert gas to a value below the limiting oxidant concentration (LOC). Values of the LOC for many gases and dusts can be found in NFPA 69 (NFPA 1997). Some commonly used inert gases used in industry are nitrogen, steam, carbon dioxide, and rare gases. Figure 3-9 shows the effect of various inert gases on the limits of flammability of methane–inert-gas–air mixtures at 25°C (77°F) and atmospheric pressure.

In the design of inerting systems one must provide sufficient inerting gas to assure not only that the normal process conditions are rendered nonflammable, but also that any credible alteration of the process environment remains outside the combustible limits. Figure 3-10 is adapted from the flammability triangular diagram presented by Zabetakis (1965) for the system methane–oxygen–nitrogen under atmospheric conditions. On this type of diagram, the sum of the three gas components is 100% at every

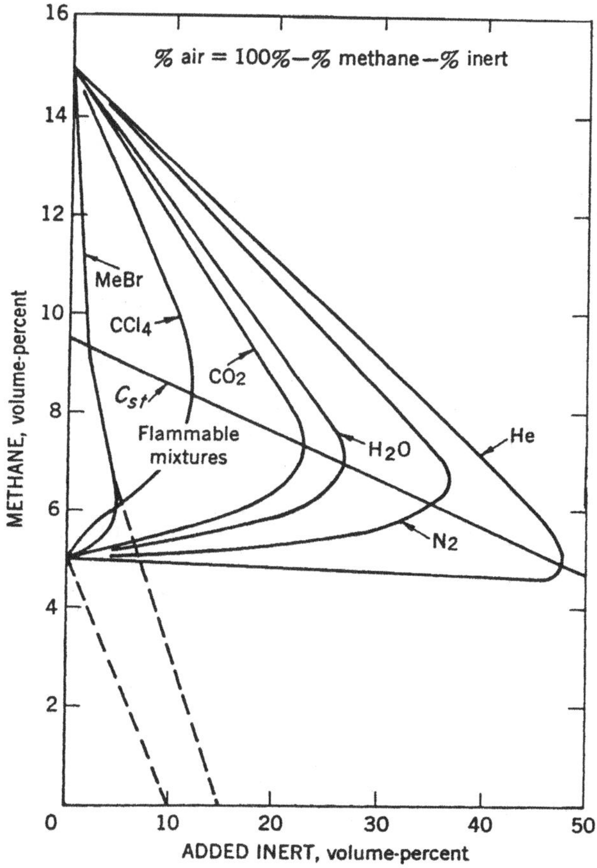

FIGURE 3-9. Effect of various gases on the flammability limits of methane–inert-gas–air mixtures at 25°C (77°F) and atmospheric pressure (Zabetakis 1965).

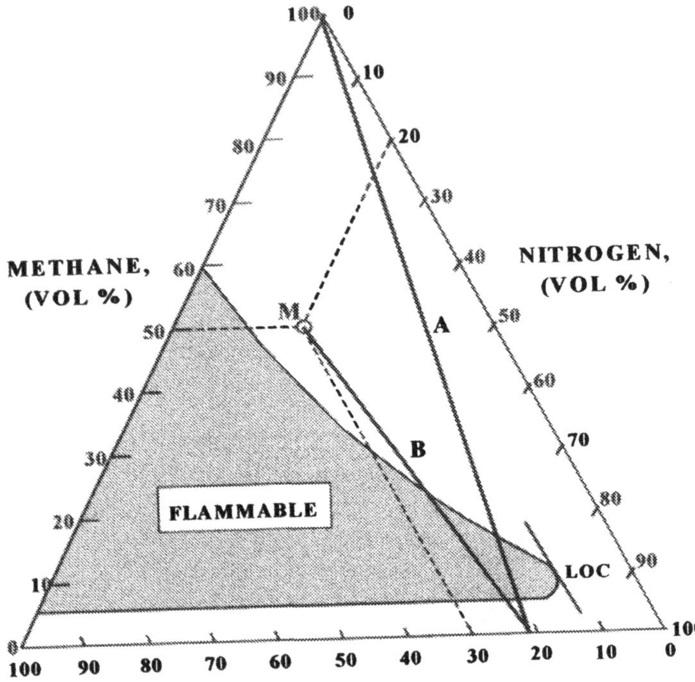

FIGURE 3-10. Flammability diagram for methane–oxygen–nitrogen mixtures. (Courtesy of L. G. Britton.)

point. On the "methane" leg of the diagram the nitrogen concentration is zero, and the flammable limits of methane in oxygen are read as 60% (UFL) and 5% (LFL). As air is added to pure methane, mixture compositions follow line "*A*," since this line represents all compositions that contain a 79:21 ratio of nitrogen to oxygen. The intersections of line "A" with the flammable envelope show that the flammable limits in air are 15% (UFL) and 5% (LFL). Now consider point "*M*," comprising 50% methane, 30% oxygen, and 20% nitrogen. This composition lies above the flammable envelope in the "methane rich" region and normally represents a safe operating condition. However, should air leak into the system, the resulting compositions follow line "*B*," which passes through the flammable envelope. Given the possibility of an air leak, available strategies are (1) operate below the LFL in air, or (2) operate so that there is insufficient oxygen to support a flame at the given concentration of fuel. In this section option (2) or "oxidant concentration reduction" is considered. There are two variants of this method:

Operation below LOC

The minimum concentration of oxygen that can support a flame is known as the "limiting oxidant concentration" or LOC, which is a singularity appearing at a fuel concentration marginally above the LFL. As shown in Figure 3-10, the LOC is identified as the line of constant oxygen concentration that is tangential to the "nose" of the flammable envelope. For methane the LOC is 12% oxygen at standard conditions. Provided the oxygen concentration is kept below the LOC, mixtures are nonflammable at all possible fuel concentrations. Therefore, where the fuel concentration is not controlled, it is common practice to control flammability by operating below the LOC.

A safety margin between the LOC and the normal oxidant concentration in the process equipment or piping system is mandated by NFPA 69 (NFPA 1997) as follows:

1. Where the oxidant concentration is continually monitored, a safety margin of at least 2 volume percent below the measured worst credible case LOC shall be maintained, unless the LOC is less than 5 volume percent, in which case, the equipment or piping shall be operated at no more than 60% of the LOC.
2. Where the oxidant concentration is not continually monitored, the oxidant concentration shall be maintained at no more than 60% of the LOC or 40% of the LOC if the LOC is below 5 volume percent. If the oxidant concentration is not continually monitored the oxidant concentration shall be checked on a regularly scheduled basis.

When inerting is used, it is prudent practice to monitor the oxidant concentration in the system by means of oxygen analyzers. The analyzers may be permanently installed, or portable ones can be used on a regularly scheduled basis. The three most commonly used oxygen analyzers are the electrochemical cell, the paramagnetic resonance type, and the zirconium catalytic cell (used for boiler and heater firing control). The zirconium catalytic cell type mounts directly in the process stream in a thermowell and does not require a sampling system. All work equally well if properly installed and maintained. Quite often a gas sample must be pretreated to remove harmful components such as water, acids, and dust to prevent damage to the analyzer cell. Also, gas sample lines should not be oversized as a long time lag can result, which would be detrimental to using the analyzer to trigger an alarm or shutdown.

A problem common to electrochemical cells is the analyzer can fail (degradation of the anode material and electrolyte or loss of sensor sensitivity), but still indicate a safe oxidant level when in reality it may not be so. Adherence to the manufacturer's recommendation as to how frequently the cell should be replaced will minimize the problem. It is possible to purchase an electrochemical cell oxygen analyzer with two cells in tandem, and an alarm to indicate when the first cell is failing. To keep oxygen analyzers operating properly, they should be calibrated and maintained on a regular schedule. In a highly hazardous system, two different types of oxygen analyzers (e.g., electrochemical cell and paramagnetic resonance), with one acting as a backup and check for the other one, can be considered. The paramagnetic resonance type exhibits the best stability and reliability for most applications. It is extremely important that a representative and timely sample be obtained for analysis. If a sample system is employed, it must be responsive and reliable. The IChemE (1983) has published guide notes on the safe application of oxygen analyzers.

It is very important that the inert gas be available from a reliable source and that the proper pressure and flow rate are always provided. A low-pressure switch and alarm are sometimes installed in the inert gas supply line to the equipment and piping. The alarm will warn the operator that a problem may be occurring with the inert gas supply. The switch is also sometimes interlocked to open up a valve in piping from a backup inert gas cylinder bank.

Following maintenance on or vessel entry into an inerted system, oxygen may be introduced via air into a single vessel or multiple vessels in a manifolded system. This may place oxygen into the manifold and create a potential for a deflagration or detonation to occur. The vessels that are opened and allow air ingress must be purged to the atmosphere so that they are below the LOC before they are reconnected to the manifold. This can be accomplished by proper valving.

Design and operating criteria for inerting systems are presented in NFPA 69 (NFPA 1997). Another excellent reference on inerting systems is provided in the Expert Commission for Safety in the Swiss Chemical Industry (ESCIS) Booklet No. 3 (ESCIS 1994).

Gas Enrichment

For nondecomposable gases the concentration of oxygen required to support a flame increases as the fuel concentration increases. For example, Figure 3-10 shows that it requires about 40% oxygen to support a methane flame in oxygen compared with only 12% oxygen at the LOC composition, which contains approximately 9% methane and 79% nitrogen. For mixtures containing increasing concentrations of methane, the oxygen concentrations required to support a flame increase along the upper bound of the flammable envelope (UFL) curve. Provided the fuel concentration is maintained sufficiently high, it is possible to operate safely at oxygen concentrations greater than the LOC. For systems in which the fuel concentration is not inherently maintained at a safe level, such as high vapor pressure liquids in closed tanks, an option is to use "gas enrichment" with a suitable fuel gas such as methane. This approach is recognized in NFPA 69.

As discussed by Britton (1996) there are sometimes advantages in operating at high concentrations of fuel rather than at low concentrations of oxygen, especially in flowing systems. The principal case considered was a methane-enriched, marine tank vent collection header as regulated by the US Coast Guard in 33 CFR 154; in certain cases the regulations require operation at less than the LOC (alarm at 8% oxygen). It was shown that designing to maintain ≥25% methane via flow control yields a larger oxygen safety factor than designing to maintain oxygen 2 volume percent below the LOC. The former approach could in some cases result in enrichment gas savings up to 50%, with attendant environmental benefits. The maintenance of at least 25% methane enrichment gas via flow control was shown to preserve nonflammability of any gas mixture not containing decomposable gases or more oxygen than could be obtained from the air. This included cases where cargo tanks were initially ballasted (inerted) with nitrogen, and cases where the UFL of the flammable cargo vapor greatly exceeded that of methane.

While the approach suggested by Britton (1996) can be applied to general in-plant vent header systems, it should be noted that marine vent collection headers continue to be regulated under 33 CFR 154. These regulations provide for the use of detonation flame arresters and other mitigating strategies for start-up, operation, and shutdown. Comparable safety systems should be considered for nonregulated systems employing gas enrichment methods.

3.5. Combustible Concentration Reduction

Combustible concentration reduction can also be used to prevent deflagrations and detonations in process equipment and piping. The combustible concentration is reduced below the lower flammable limit (LFL) by means of ventilation (air dilution).

NFPA 69 (NFPA 1997) contains information on basic design considerations, design and operating requirements, and instrumentation requirements. Appendix D presents methods for ventilation calculations, including the time required for ventilation to reduce the concentration to a safe limit, the number of air changes required for reaching a desired combustible concentration, and the time required to reach a buildup of combustible concentration in an enclosed volume.

NFPA 69 stipulates that the combustible concentration be maintained at or below 25% of the LFL with the following exceptions:

1. When automatic instrumentation with safety interlocks is provided, the combustible concentration may be maintained at or below 60% of the LFL.
2. In aluminum powder production systems, the combustible concentration is permitted to be maintained at or below 50% of the LFL.

Under certain situations, for specific equipment configurations, it may be possible to safely operate above the NFPA limit of 60% of the LFL. If this method of operation is to be considered, system-specific test data should be generated which demonstrates that the combustible concentration can be controlled in a safe manner, and only then in consultation with appropriate company and (where required) regulatory authorities.

3.6. Deflagration Suppression

In many cases, deflagration flames can be extinguished before unacceptable pressures occur within process equipment and piping if the onset of combustion can be detected early and an appropriate extinguishing agent can be delivered to the proper location within equipment or piping. The technique of deflagration suppression is applicable to most flammable gases and vapors, combustible mists, or combustible dusts that are subject to deflagration in a gas phase oxidant. Suppression systems are active systems that include components for detection, suppressant delivery, electrical supervision to assure readiness to operate, and interlock functions to shut off or isolate other process equipment connected to the equipment to be protected.

Deflagration suppression systems can be applied to a large number of types of process equipment, rooms, and piping systems (including vent manifolded systems) (NFPA 69 1997). Deflagration suppression is a competitive process between a rising rate of combustion heat release and a delayed, but rapid, delivery of extinguishing agent. The deflagration will be suppressed when the unburned fuel–air mixture has been rendered noncombustible due to the addition of an extinguishing agent, or the combustion zone has been cooled to the point of extinguishment, or the reaction kinetics are impeded. The time required for a suppression system to stop a flame front from propagating is dependent on the equipment or piping volume (time increases with increasing volume) and the flame speed of the material being handled (time decreases with increasing flame speed). For example, the time to suppress a deflagration in a 1.9 cubic meter vessel takes about 100 milliseconds, while it takes 250 milliseconds for the same event in a 25 cubic meter vessel (CCPS 1993). Each application requires experimental validation of suppression system design.

The sequence of events that occurs in a suppressed explosion is shown in Figure 3-11 for a pressure-threshold-type detector. For rate of pressure rise and optical flame detectors the sequence differs slightly. Figure 3-11 is for a dust explosion, but the sequence is similar for a gas explosion. Referring to Figure 3-11, after ignition the pressure in the equipment or piping rapidly increases. After a time t_d the explosion pressure somewhere in the equipment reaches the pressure P_a (at which the pressure detector is set), commonly about 0.05 bar. The explosion pressure wave propagates through the vessel at the speed of sound so that it takes a finite time t_e before the pressure threshold is attained at the explosion detector. This equalization time depends on the relative positions of the detector and the ignition source. In the case of a cubical vessel with 3-meter sides, and the detector in the center of one side and the ignition source in the opposite corner, the equalization time is about 12 milliseconds. The time actually taken to fire the suppressor actuator is usually quite small (in the order of a few milliseconds). The suppressant is then injected across the vessel initially at the discharge velocity of the suppressor (about 40 m/s) and thus it takes a finite time for the suppressant to reach all parts of the vessel. The time required for the suppressant to reach the furthest part of the vessel can be considerably reduced by the use of more than one suppressor. In Figure 3-11, this time is labeled t_s.

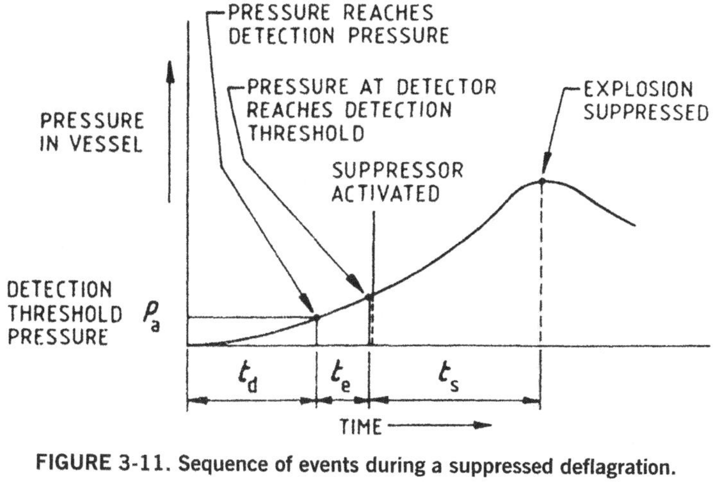

FIGURE 3-11. Sequence of events during a suppressed deflagration.

As mentioned earlier, a suppression system consists of three subsystems for (a) detection, (b) extinguishment, and (c) control and supervision. Incipient deflagrations are detected using either pressure detectors, rate of pressure rise (or "rate") detectors, or optical flame detectors. Pressure detectors are employed in closed process equipment or piping, and particularly where dusty atmospheres are present. Rate detectors are used in processes that operate at pressures significantly above or below atmospheric pressure. Optical detectors may be infrared (IR), ultraviolet (UV) or hybrid (i.e., both IR and UV) depending on the flame to be detected and the absorbent properties of the operating environment.

The extinguishment subsystem consists of one or more high rate discharge (HRD) suppressant containers charged with suppressing agent and propellant. Normally, dry nitrogen is used to propel the agent out of the container into the equipment or piping. The propellant pressure is normally in the range of 300 to 900 psig, depending on the supplier of the suppression system. An explosive charge is electrically detonated and opens valves providing rapid agent delivery to the equipment or piping being protected.

Common extinguishing agents are water, Halon substitutes, and dry chemical formulations typically based on sodium bicarbonate or ammonium dihydrogen phosphate. The extinguishing mechanism of each agent is often a combination of thermal quenching and chemical inhibition. The selection of the appropriate agent is usually based on several considerations such as effectiveness, toxicity, cost, product compatibility, and volatility. Water is often a very effective suppressant, and should be used whenever possible since it is not toxic and is easier to clean up in comparison to the other types of suppressants. Dry chemical agents have been used for many years in Europe and are being used now more often in the United States. Halons were used for many years as they were very effective suppressants, but they have been outlawed in many countries because of their adverse effect on stratospheric ozone (the "ozone layer"). Numerous substitutes are now available, but none have been found to be as effective as Halon 1301 and other Halons. One new substitute that is being widely used is FM 200™ (a hydrofluorocarbon).

The control and supervision of a suppression system is achieved using an electrical power/control unit with 24-hour (minimum) standby battery backup power. The unit supplies sufficient energy to accomplish the following (NFPA 69 1997):

- power all detection devices
- energize all electrically fired initiators
- energize visual and audible alarms

- transfer all auxiliary control and alarm contacts
- control system disabling interlock and process shutdown circuits

Control and supervision systems should be designed with circuit monitoring and self-diagnostic testing to verify that the field sensors and devices are electrically active and connected. The system should alarm when an electrical fault is detected.

It should be recognized that deflagration suppression systems have a number of shortcomings, such as:

- spurious activation (false trips)
- servicing problems (testing, bypassing for maintenance)
- clean-up after activation for a real event
- good for only one event and then they have to be refilled
- potential for overpressure of the protected equipment upon discharge which needs to be considered in design

Figure 3-12 is a schematic of a deflagration suppressant system for process equipment. Each application requires experimental validation of the suppression system design.

Suppressant systems for piping are called suppressant or chemical barriers and are discussed in Section 3.8.

Additional information on deflagration suppression systems can be found in NFPA 69 (1997), CCPS Guidelines (1993), Bartknecht (1989), Schofield and Abbott (1988), and Eckhoff (1997).

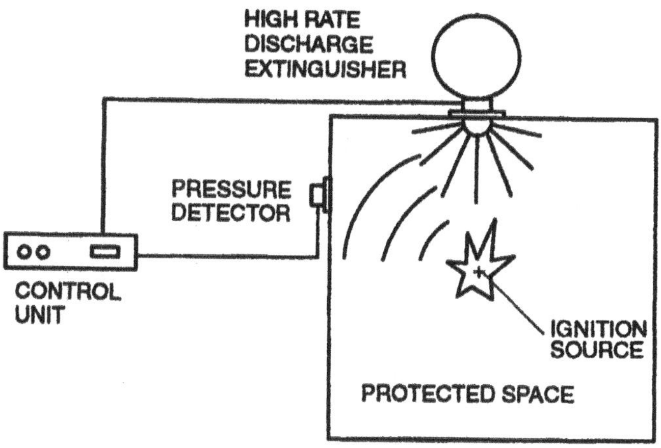

FIGURE 3-12. Schematic of a deflagration suppression system for process equipment.

3.7. Deflagration Pressure Containment

Deflagration pressure containment is an approach for selecting the design pressure of a vessel so that it is capable of withstanding the maximum pressure resulting from an internal deflagration. Vessels or process equipment can be designed to either

- prevent rupture, but allow deformation (known as "shock-resistant" design in Europe), or
- prevent rupture and deformation (requires a thicker vessel wall).

NFPA 69 (1997) provides equations for calculating the required design pressures for both types of containment design. It also discusses the limitations of deflagration pressure containment design.

In Europe process equipment such as spray dryers, fluid-bed dryers, and mills are available in "shock-resistant" designs for pressures up to 10 barg (145 psig).

Pressure containment can also be provided by using piping systems with a pressure rating above the anticipated maximum pressure generated during a deflagration.

More information about pressure containment design is available in NFPA 69 and the books by Bartknecht (1981, 1989) and Eckhoff (1997).

3.8. Equipment and Piping Isolation

It is common practice in the chemical process industries to provide isolation devices for stopping flame fronts, deflagration pressures, pressure piling, and flame-jet ignition between process equipment interconnected by pipes or ducts. There are several devices for providing this isolation as follows:

- suppressant barriers
- fast-acting valves
- material chokes
- flame front diverters

These are discussed below.

Suppressant Barriers

This type of isolation device (also called a chemical barrier) is similar to deflagration suppression systems used on process equipment. This barrier system consists of an optical sensor, installed in the pipeline or duct between two items of equipment, that detects an oncoming deflagration flame and emits a signal to a control unit. The amplified signal triggers the detonator-activated valve in a suppressant bottle, which injects an extinguishing agent into the pipeline through suitable nozzles. Pressure sensors are not normally used for pipeline barriers since there is no clear correlation between the front of the pressure wave and the flame front, and pressure sensor response times often are too slow for use in this application.

Suppressant barrier systems have the same shortcomings as cited in Section 3.6. In addition, the location of the sensor is critical to the successful isolation of a deflagration flame in a piping system. Bartknecht (1981) states that the flame sensor is installed at a distance of 1 meter from the ignition source, and the extinguisher nozzles at a distance of 10 meters from the ignition source.

Figure 3-13 is a schematic of a deflagration suppressant barrier system for pipelines.

Further information on suppressant barriers can be found in NFPA 69 (1997), and the books by Bartknecht (1981 and 1989) and Eckhoff (1997).

Fast-Acting Valves

A variety of fast-acting valves are available, including slide gate, flap (butterfly), and float (poppet) valves. Slide gate and flap type valves are actuated (closed) upon a signal from a detector (sensor) in the pipeline between two items of interconnected process equipment. The detector sends a

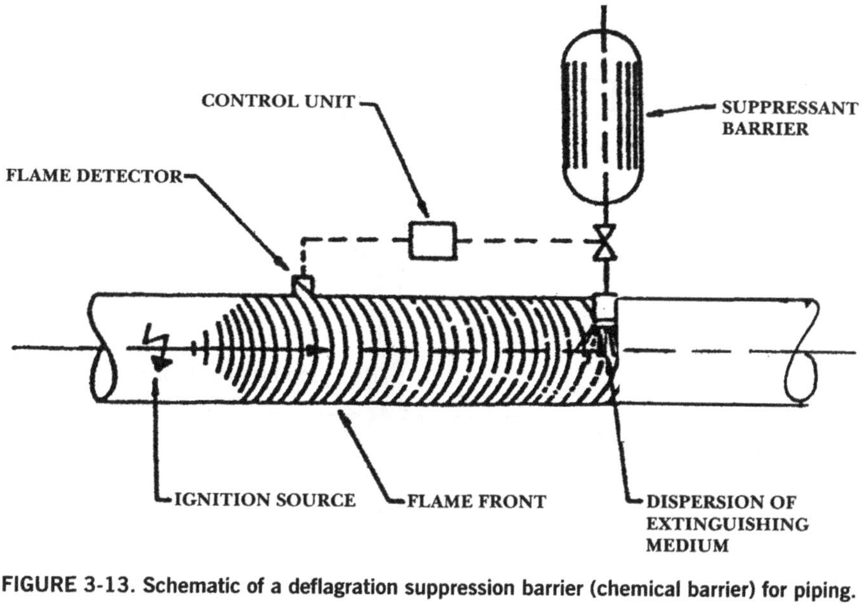

FIGURE 3-13. Schematic of a deflagration suppression barrier (chemical barrier) for piping.

signal to a compressed air cylinder which then discharges the compressed air to a mechanism at the top of the valve, thereby closing the valve. A typical closing time for a fast-acting valve is about 25 milliseconds. The deflagration detector is located about 1 meter away from the source of ignition (equipment), and the fast-acting valve is usually installed 5 to 10 meters along the connecting pipeline, so that by the time the flame front reaches it, the valve is fully closed. A fast-acting slide gate isolation valve is shown in Figure 3-14 and a fast-acting flap (butterfly) valve is shown in Figure 3-15.

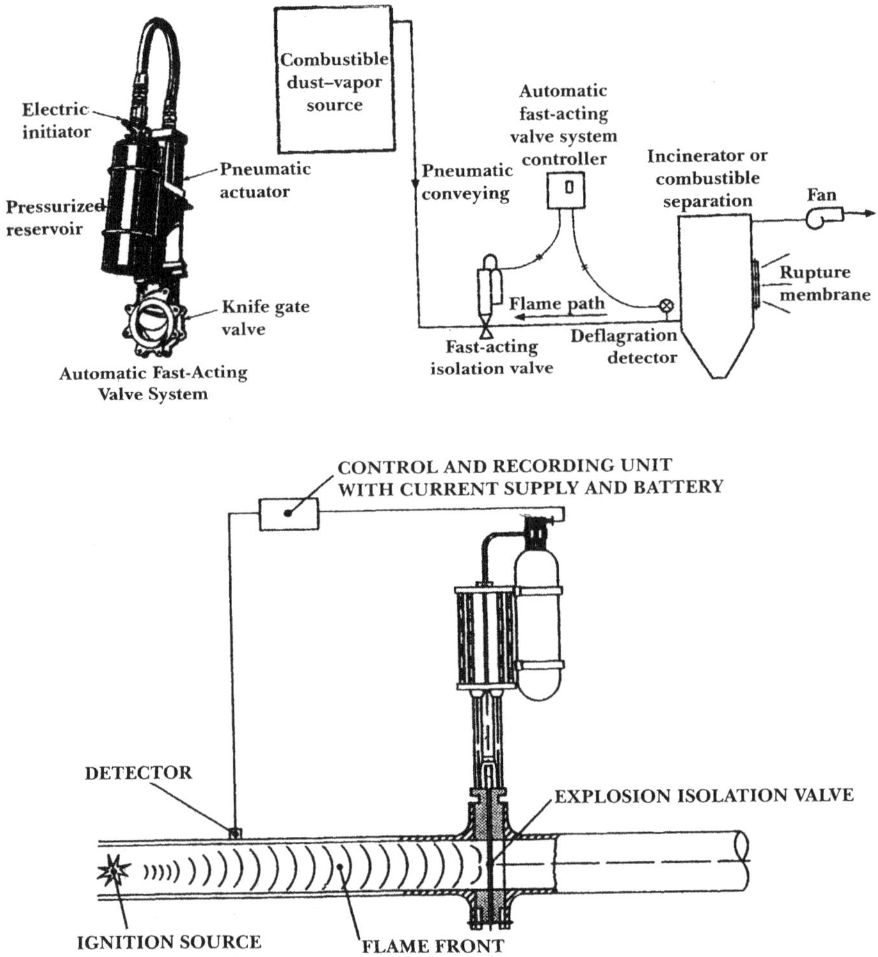

FIGURE 3-14. Fast-acting slide gate valve isolation system (NFPA 69).

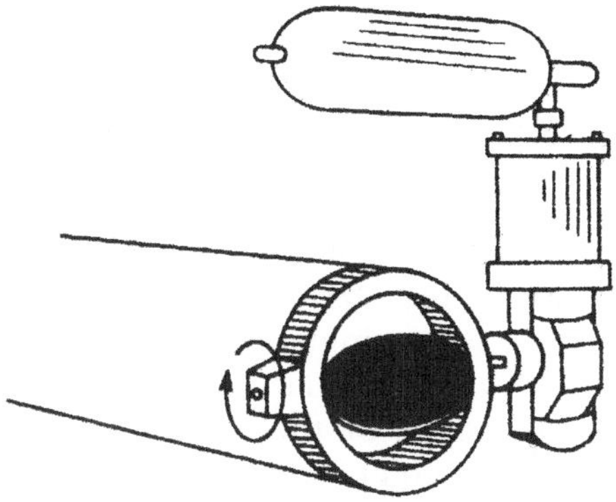

FIGURE 3-15. Fast-acting flap (butterfly) isolation valve (Eckhoff 1997).

Another type of fast-acting valve is the Ventex™ valve (a float type valve), which is activated by the deflagration pressure wave advancing through the pipeline. This valve must be installed horizontally, and its minimum activation pressure is in the order of 0.1 barg (1.45 psig). In normal operation, the gas or dust being conveyed in the pipeline flows around the valve poppet without causing any significant off-set as long as the flow velocity is less than about 20 m/s. However, in the case of a deflagration in the pipeline, the advancing pressure wave pushes the valve poppet in the axial direction until it hits a neoprene gasket, where it is held in position by a mechanical catch lock, which can be released from the outside. However, the neoprene gasket may be adversely affected in high temperature environments. Figure 3-16 shows a schematic drawing of a Ventex™ valve.

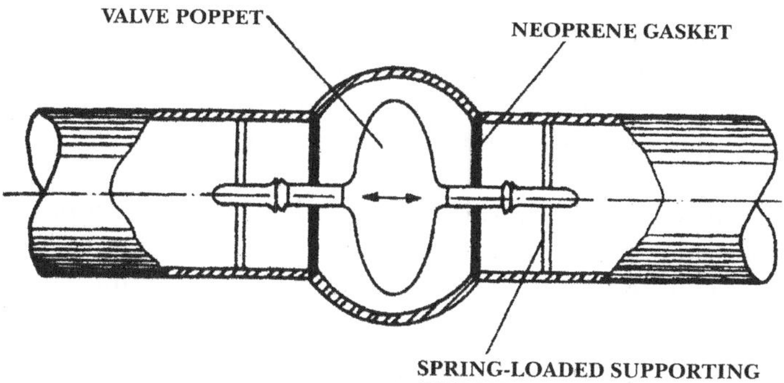

FIGURE 3-16. Ventex™ (float type) fast-acting isolation valve (Eckhoff 1997).

Another, more recent, type of fast-acting valve is the Exkop™ valve. The valve trim (internals) is actually a rubber bladder surrounded by an air chamber. In the event of a deflagration, an electrical signal is sent from a sensor, typically mounted on a deflagration relief device, to the Exkop™ valve air tank mounted integral to the valve. The air tank discharges air to the chamber surrounding the rubber bladder and compresses it, which pinches off flow in the pipeline. Because the rubber bladder has relatively little mass, it is both very fast acting and imparts low shock to the piping. The rubber bladder, however, may be adversely affected by high temperatures. One advantage of the Exkop™ valve is that it can be mounted fairly close to the equipment in which the deflagration occurs, usually 15 to 18 feet away from the equipment. It also immediately rearms itself with ordinary plant air and is automatically placed back in service. Figure 3-17 shows details of an Exkop™ valve installation. Exkop™ valves are only used for isolation of piping carrying a dust–air mixture. They are not recommended for stopping gas deflagrations as tests have shown that in a gas/vapor deflagration, the flame propagation is too fast for the valve to work effectively (Stevenson 2000).

Material Chokes

Flame propagation can also be stopped between process equipment handling bulk solids and powders by judicious selection and design of bulk solids/powders conveying equipment such as rotary valves (rotary airlocks) and screw conveyors. The mass of bulk solids/powders contained in these items of equipment provide a tortuous path through which the gas and flame have to pass, and so can act as a "material choke" when certain design features are implemented.

Rotary valves will generally prevent flame propagation if the following criteria are followed (Bartknecht 1989, NFPA 69 1997):

- two vanes per side are always in contact with the housing
- the vanes or tips are made out of metal (no plastic vanes) and
- the gap between the rotor and housing is ≤0.2 mm

In screw conveyors the removal of part of the screw will ensure that a plug of bulk solids/powder will always remain as a choke (Eckhoff 1997).

Several considerations have to be taken into account when using a rotary valve as a material choke. When a deflagration occurs in the equipment upstream of the rotary valve, the rotary valve has to be stopped immediately by a suitable detector in order not to pass burning or glowing solids into downstream equipment, which could then cause a second fire or

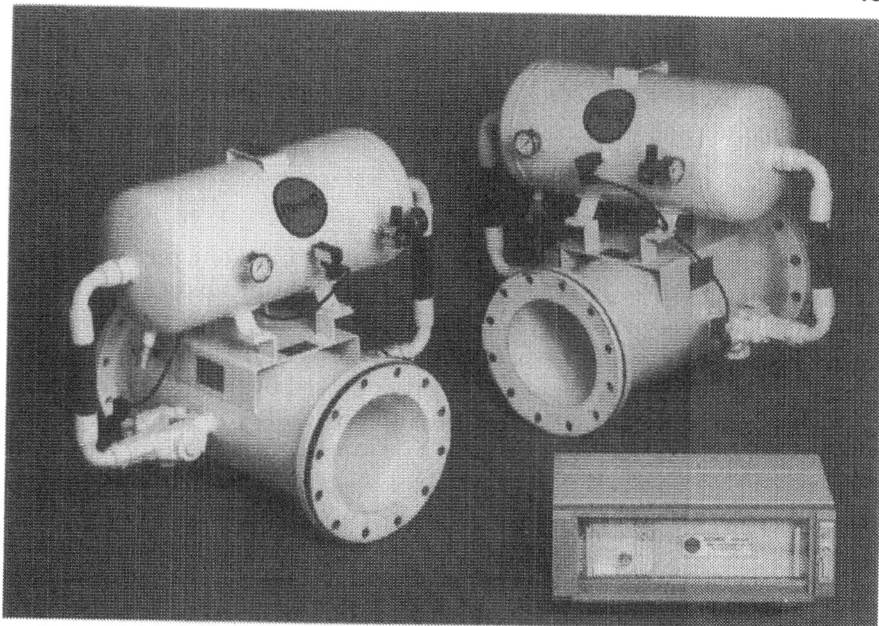

(a) Photo of Exkop fast-acting isolation valve

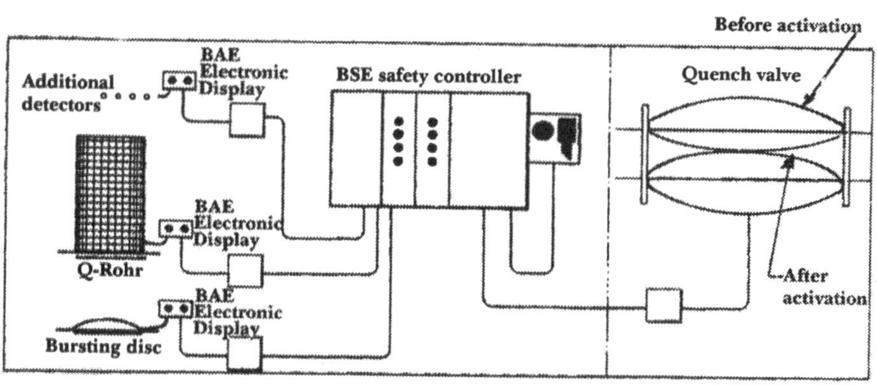

(b) Schematic of Exkop valve before and after activation

FIGURE 3-17. Exkop™ fast-acting isolation valve (quench valve). (Source: Cv Technology, Inc.)

deflagration. Rotary valves must be tested for their suitability as flame arresters as well as for their pressure rating with appropriate explosion tests (Bartknecht 1989). These devices must be properly maintained to ensure that normal wear and tear does not result in a loss of seal between the rotor blades and the housing.

Flame Front Diverters

Flames can also be prevented from propagating from one piece of equipment to another through interconnecting piping by means of a flame front diverter. The basic principle of operation of this device is that the deflagration is vented at a point where the flow direction is changed by 180 degrees. Due to the inertia of the fast flow caused by the deflagration, the flow will tend to maintain its direction upward rather than making a 180 degree turn when the velocity is low (at normal conditions). When the high-speed deflagration flame continues upward, it pushes open either a hinged cover or bursts a rupture disk located at the top of the diverter, allowing the flame to be released to the atmosphere. The location of a flame front diverter must be chosen so that the release of the flame does pose a hazard to people or equipment.

Some flame front diverters have demonstrated the ability to successfully divert deflagration flames by directing them to the atmosphere. However, in some cases, tests have indicated that some diverters have been ineffective in completely diverting a deflagration; but where this has occurred, the deflagration severity has been reduced (NFPA 69 1997). Therefore, before they are used, it is recommended that they be tested for the desired application. Figure 3-18 shows several flame front diverter designs.

More information about flame front diverters is presented in NFPA 69 (1997) and books by Bartknecht (1989) and Eckhoff (1997).

3.9. References

3.9.1. Regulations, Codes of Practice, and Industry Standards

CSA (Canadian Standards Association) CSA-Z343. 1998. *Test Methods for In-Line and Firebox Flame Arresters*. Canadian Standards Association, Toronto, Ontario, Canada.

NFPA 68. 1998. *Guide for Venting of Deflagrations*. National Fire Protection Association, Quincy, MA.

NFPA 69. 1997. *Standard on Explosion Prevention Systems*. National Fire Protection Association, Quincy, MA.

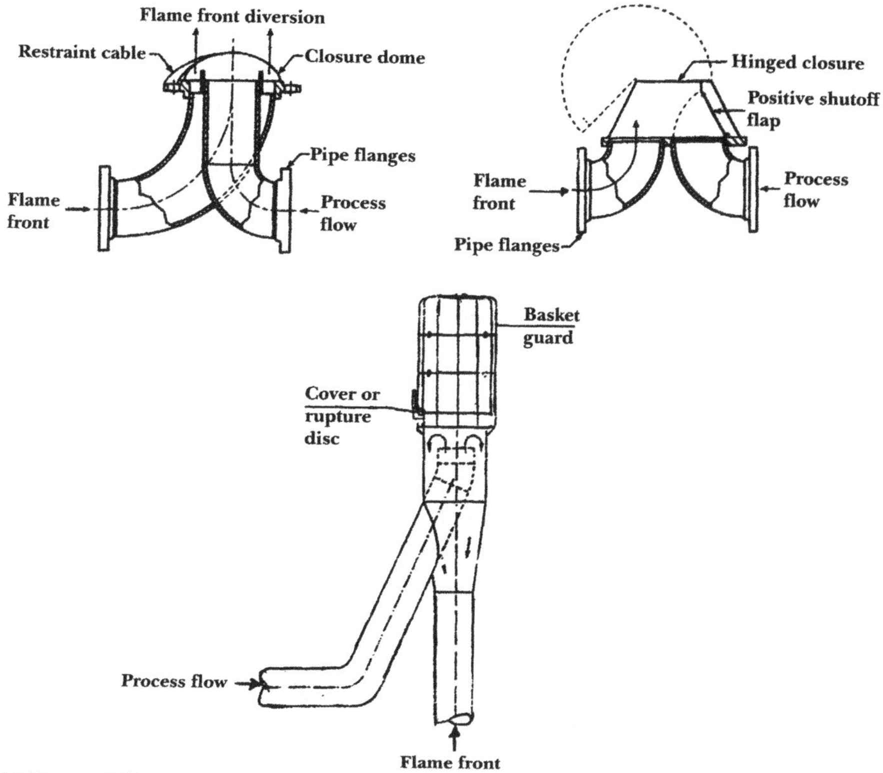

(a) Types of flame front diverters

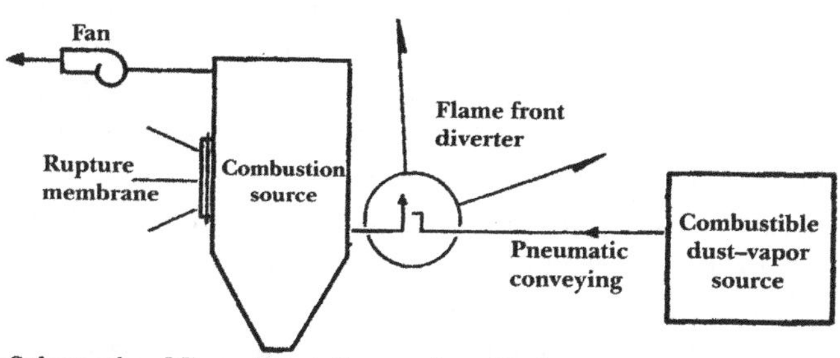

(b) Schematic of flame front diverter installation

FIGURE 3-18. Flame front diverters (NFPA 69, Bartknecth 1989).

3.9.2. Specific References

Bartknecht, W. 1981. *Explosions: Course, Prevention, Protection.* Springer-Verlag, New York, NY.

Bartknecht, W. 1989. *Dust Explosions: Course, Prevention, Protection.* Springer-Verlag, New York, NY.

Britton, L. G. 1996. Operating Atmospheric Vent Collection Headers Using Methane Gas Enrichment. *Process Safety Progress,* 15(4), 194–212.

CCPS. 1993. *Guidelines for Engineering Design for Process Safety.*American Institute of Chemical Engineers, Center for Chemical Process Safety, New York, NY.

Chatrathi, K. 2001. Personal communication from Kris Chatrathi, Fike Corporation, to S. S. Grossel, October 15, 2001.

Eckhoff, R. K. 1997. *Dust Explosions in the Process Industries.* 2nd ed. Butterworth-Heinemann, Boston, MA.

ESCIS. 1994. *Inerting.* Booklet No. 3, Expert Commission for Safety in the Swiss Chemical Industry, Basel, Switzerland.

Halstrick, V. 1995. Technical Report Part I, Protego Fundamentals. Braunschweiger Flammenfilter GmbH, Braunschweig, Germany.

Howard, W. B. 1982. Flame Arresters and Flashback Preventers. *Plant/Operations Progress,* 1(4), 203–208; 2(1), 71–72 (January 1983).

IChemE. 1983. *Guide Notes on the Safe Application of Oxygen Analyzers.* The Institution of Chemical Engineers, Rugby, Warwickshire, UK.

Knittel, T. 1993. In Line (Detonation) Flame Arresters—Function, Certification, Selection and Application. Paper 10c. Paper presented at the 27th Annual AIChE Loss Prevention Symposium, March 30–April 1, 1993, Houston, TX.

Roussakis, N. and Lapp, L. 1991. A Comprehensive Test Method for Inline Flame Arresters. *Plant/Operations Progress,* 10(2), 85–92 (April).

Schofield, C. and Abbott, J. A. 1988. *Guide to Dust Explosion Prevention and Protection,* Vol. 2: *Ignition Prevention, Containment, Inerting, Suppression and Isolation.* Institution of Chemical Engineers, Rugby, Warwickshire, UK.

Stevenson III, W. J. 1998. Dust Explosion Mitigation Using Q-Rohr and Exkop. 32nd Annual AIChE Loss Prevention Symposium, Paper 3f. American Institute of Chemical Engineers, New York, NY.

Stevenson III, W. J. 2000. Personal communication from W. J. Stevenson III, Cv Technology, Inc. to S. S. Grossel (January 31, 2000 and February 14, 2000).

Westech Industrial Ltd. 1989. Flame Arrester Seminar Notes. Westech Industrial Ltd., Calgary, Alberta, Canada.

White, R. E. and Oswald, C. J. 1992. Mitigation of Explosion Hazards of Marine Vapor Control Systems. Southwest Research Institute, Final Report, SWRI Project No. 06-4116 for the American Petroleum Institute, Washington, DC (October 1992).

Zabetakis, M. G. 1965. *Flammable Characteristics of Combustible Gases and Vapors. Bureau of Mines* Bulletin 627. U.S. Department of the Interior, Washington, DC.

4

Overview of Combustion and Flame Propagation Phenomena Related to DDAs

4.1. Introduction to the Chemistry and Physics of Flame Propagation

This chapter presents fundamental concepts of combustion theory relating both to flame propagation and DDA technology.

Flame propagation is driven by the energy released from an exothermic reaction. This energy release can be the result of combustion, exothermic decomposition, or (in some cases) a combination of the two. Respective examples are combustion of methane in air, decomposition of acetylene, and combustion of rich fuel mixtures containing decomposable gases. DDA design requires a basic understanding of flame propagation in the gas phase, specifically where a gas mixture ignited in equipment propagates a deflagration or detonation flame through a pipe. The DDA is primarily an isolation device, preventing flame propagation from one location to another. Depending on the type of DDA and the application, reduction of peak overpressures and other benefits may be gained from their use.

The processes by which energy is released from nonstationary flames involve complex interactions between chemistry and fluid dynamics. A good introduction to combustion is presented in the books by Baker et al. (1983), Bradley (1969), Glassman (1996), Griffiths and Barnard (1995),

Lewis and von Elbe (1987), Strahle (1993), Stull (1977), and Van Dolah and Burgess (1974). The following sections provide an elementary overview.

4.1.1 Combustion Chemistry and Thermodynamics

Reaction Stoichiometry

The chemical reactions that occur in flames transform an initial reactant mixture into final reaction products. In the case of fuel-oxygen combustion, the final products are principally water vapor and carbon dioxide, although numerous other products such as carbon monoxide may be formed, depending on the reactant composition and other factors. If the ratio of fuel-to-oxygen is "stoichiometric," the final reaction products, by definition, contain no excess fuel or oxygen. Theoretically, this means that partial oxidation products such as CO (itself a fuel) are not formed. In reality, partial oxidation products such as CO or OH are formed by high temperature reactions. For example, the molar stoichiometric reaction of methane is written:

In oxygen:

$$CH_4 + 2O_2 \rightarrow CO_2 + 2H_2O \tag{4-1a}$$

In air:

$$CH_4 + 2O_2 + 2(79/21)N_2 \rightarrow CO_2 + 2H_2O + 2(79/21)N_2 \tag{4-1b}$$

In practice, nitrogen reacts with oxygen at high temperatures, forming small amounts of nitric oxide and other species. However, this is neglected when writing the stoichiometric equation and the fuel is always assumed to react simply with oxygen. Nitrogen is considered an "inert diluent." However, the presence of this and other inert gases must be included when defining the "stoichiometry" of a mixture.

An important parameter in combustion science is the "stoichiometric ratio," represented by the symbol "φ." This is the concentration of fuel expressed as a fraction of the stoichiometric concentration of fuel, both on a molar basis. This nondimensional quantity is useful in describing mixtures; if the ratio $\varphi < 1$, it is a lean mixture, and if $\varphi > 1$, it is a rich mixture. Combustion becomes more rapid as φ approaches unity. Fuels that diffuse slower than oxygen, such as ethane and higher paraffins, exhibit faster burning and easiest ignition at slightly rich compositions ($\varphi \sim 1.5$). Conversely, fuels that diffuse faster than oxygen, such as methane and hydrogen, exhibit this behavior at slightly lean compositions. The mixtures used during the certification of DDAs in accordance with the USCG protocol

TABLE 4-1
Comparison of Stoichiometric and Most Easily Ignitable Concentrations for Five Gases and Vapors in Air

Gas/Vapor	Stoichiometric Concentration (vol %)	Most Easily Ignitable Concentration (vol %)
Ethylene	6.53	7.5
Hexane	2.16	3.8
Hydrogen	29.5	28.0
Methane	9.47	8.5
Propane	4.02	5.2

(33CFR Part 154, Appendix A) correspond to the "most easily ignitable" composition (lowest MIE). This is usually known as the optimum concentration. Table 4-1 shows a comparison between the stoichiomeric and the most easily ignitable concentrations of five gases and vapors that are used in DDA certification tests.

From Eq. (4-1a) the stoichiometric concentration of methane in oxygen is 1 part in 3 = 33.3 mole percent methane. From Eq. (4-1b) the approximate stoichiometric concentration of methane in air is 1 part in 3 + (158/21) = 9.5 mole percent methane. Thus, a mixture of 15 mole percent methane in oxygen has a stoichiometric ratio φ = 15/33.3 = 0.45 (lean), while the same methane concentration in air has a stoichiometric ratio φ = 15/9.5 = 1.58 (rich).

For paraffins the stoichiometric ratio decreases as the number of carbon atoms increases. Using a more precise calculation (which includes other species such as CO, OH, etc.) than that shown in Eq. (4-1b), methane's stoichiometric ratio in air is 9.48 mole percent, propane's is 4.01 mole percent, and hexane's is 2.16 mole percent. Hydrogen, which combines with oxygen to form only water, has a stoichiometric ratio of 29.6 mole percent in air.

Enthalpy and Heat of Reaction

The amount of energy that can be released from a given chemical reaction is determined from the energies (enthalpies of formation) of the individual reactants and products. Enthalpies are usually given for substances in their standard states, which are the stable states of pure substances at atmospheric pressure and at 25°C. The overall heat of reaction is the difference between the sums of the standard enthalpies of formation of the products

TABLE 4–2

Heats (Enthalpies) of Formation of Selected Chemicals at 25°C (Glassman 1996)

Chemical	Name	State	ΔH_f^c (kJ/mol)	Δh_f^c (kJ/g)
C	Carbon	Vapor	716.67	59.72
N	Nitrogen atom	Gas	472.68	33.76
O	Oxygen atom	Gas	249.17	15.57
C_2H_2	Acetylene	Gas	227.06	8.79
H	Hydrogen atom	Gas	218.00	218.00
O_3	Ozone	Gas	142.67	2.97
NO	Nitric oxide	Gas	90.29	3.01
C_6H_6	Benzene	Gas	82.96	1.06
C_6H_6	Benzene	Liquid	49.06	0.63
C_2H_4	Ethene	Gas	52.38	1.87
N_2H_4	Hydrazine	Liquid	50.63	1.58
OH	Hydroxyl radical	Gas	38.99	2.29
O_2	Oxygen	Gas	0	0
N_2	Nitrogen	Gas	0	0
H_2	Hydrogen	Gas	0	0
C	Carbon	Solid	0	0
NH3	Ammonia	Gas	–45.90	–2.70
C2H4O	Ethylene oxide	Gas	–51.08	–0.86
CH4	Methane	Gas	–74.87	–4.68
C2H6	Ethane	Gas	–84.81	–2.83
CO	Carbon monoxide	Gas	–110.53	–3.95
C_4H_{10}	Butane	Gas	–124.90	–2.15
CH_3OH	Methanol	Gas	–201.54	–6.30
CH_3OH	Methanol	Liquid	–239.00	–7.47
H_2O	Water	Gas	–241.83	–13.44
C_8H_{18}	Octane	Liquid	–250.31	–0.46
H_2O	Water	Liquid	–285.10	–15.84
SO_2	Sulfur dioxide	Gas	–296.84	–4.64
$C_{12}H_{16}$	Dodecane	Liquid	–347.77	–2.17
CO_2	Carbon dioxide	Gas	–393.52	–8.94
SO_3	Sulfur trioxide	Gas	–395.77	–4.95

and the reactants. Table 4-2 lists enthalpies of formation (also called heats of formation) for a number of representative chemicals (Glassman 1996). Heats of formation for many chemicals are listed *in Perry's Chemical Engineers' Handbook* (Green and Maloney 1997).

Flame Temperature

If no heat is ultimately lost to the surroundings, all of the energy released by a flame raises the temperature of the reaction products, and the final temperature is called the adiabatic flame temperature. The adiabatic flame temperature can be calculated with the assumption that the reaction products achieve chemical equilibrium at the calculated temperature, which is sometimes denoted as "CART" (calculated adiabatic reaction temperature). There are two general cases.

The first case is where a flame travels through an open space (expanding volume) so that the pressure remains constant. The "constant pressure" CART is most characteristic of the flame itself. All flames must initially propagate at constant pressure, so this parameter determines the ability of a particular mixture to propagate a flame. It often closely approximates the temperature attained in a thin zone known as the "reaction zone" of the flame. In practice, chemical equilibrium is not always attained and additional complications arise in high temperature flames, where dissociation of the products depresses the CART.

The second case is where the flame travels in a closed space (constant volume); the pressure increases constantly during flame propagation and this in turn compresses the reaction products, increasing their temperature. Thus, the "constant volume" CART is greater than the "constant pressure" CART. The importance of the "constant volume" CART is that it allows the final pressure to be calculated for flame propagation inside unvented enclosures. For most stoichiometric fuel (most hydrocarbon and organic chemical gases and vapors) mixtures with air the final pressure is in the range 8–12 times the initial absolute pressure. This value typically decreases to 3–5 times the initial absolute pressure near the flammability limits. This is because the pressure ratio is the product of the absolute temperature ratio and the molar expansion ratio; since molar expansion ratios are often close to unity, and flame temperatures decrease to about 1000–1500°K close to the flammability limits, the temperature ratio is typically in the range of 3–5.

Reaction Chemistry and Chemical Kinetics

Chemical reactions seldom proceed at constant temperature and pressure; neither do the reactants transform instantaneously into products. From a thermodynamic viewpoint it is only the initial and final states that are of concern. However, from a chemistry viewpoint, distinct and complex chemical reaction pathways exist, involving other intermediate chemical states described by the detailed chemical kinetics of the reaction. For example, in the global reactions described in Eqs. (4-1a) and (4-1b), not only must the destruction or creation of CH_4, O_2, H_2O, and CO_2 be considered, but other molecules and radicals such as CH_3, CH, CO, OH, and HO_2 should also be included. Whether a flame propagates, depends not only on the magnitude of the chemical energy released and the adiabatic flame temperature, but also on the rate at which the reaction proceeds. While all combustion processes depend on the total amount of energy released by chemical reaction, not all depend on the rate of reaction, provided that it exceeds some minimum value. Detonation waves are dominated by pure thermal autoignition chemistry and kinetics. A diffusion flame, as a premixed flame is also dependent on chemistry, with the kinetics modulated by heat diffusion that changes the temperature and diffusion of reactive species from the main exothermic reaction zone. In general, chemical reaction rates increase exponentially with temperature.

4.1.2. Flammability Characteristics

The discussion below pertains to various aspects of flammability that are relevant to understanding flame propagation.

Flammable Limits

Under certain conditions flames can no longer propagate in a self-sustaining manner. If the ratio of fuel to oxygen is varied away from the stoichiometric values or the degree of dilution by an inert gas is increased, the temperature of the products (the adiabatic flame temperature) decreases to a point where the reactions can no longer be sustained. The two main limits of interest occur where there is a deficit or surplus of fuel, termed the lower (lean) and upper (rich) flammable limits (abbreviated LFL and UFL). The LFL and UFL are normally determined for mixtures of fuel in air at ambient temperature and pressure, but values are also available for other oxidants (Coward and Jones 1952). For hydrogen in air, experiments have shown that these are about 4 volume percent and 75 volume percent hydrogen, respectively. The corresponding values for methane–air are 5 volume percent and 15 volume percent methane, respectively.

Flammable limits are affected by both temperature and pressure. In general, as the temperature of a gas mixture increases, the size of the flammability zone increases. For mixtures in air, the UFL increases and the

LFL decreases, which broadens the range over which the mixture is flammable. In general, pressure has little effect on the LFL, except at low pressures (e.g., about 100 mmHg for methane) where combustion is not possible. As the pressure increases, the UFL generally increases, broadening the flammability range. Equations are available for calculating the effect of temperature on the LFL and UFL, and pressure on the UFL (Coward and Jones 1952, Zabetakis 1965).

Both LFL and UFL values for mixtures can be estimated by use of the Le Chatelier equation (Crowl and Louvar 1990). However, the methods have some limitations with respect to calculating the UFL for certain mixtures. Britton (1996) determined that the equation does not apply to the UFL of mixtures containing decomposable components such as ethylene oxide or to mixtures containing ethyl ether. Mashuga and Crowl (2000) discuss the derivation of Le Chatelier's mixing rule for flammable limits.

Flammable limits are important as they indicate the range of concentrations within which a combustion reaction may occur. If a concentration of a fuel–oxidant mixture can be maintained below the LFL or above the UFL, then there is no possibility of flame propagation. Figure 3-10 (page 32) is a typical flammability diagram with the flammable zone between the LFL and the UFL indicated.

Limiting Oxidant Concentration (LOC)

In order to propagate a flame, there must be an oxygen concentration above a specific minimum value known as the limiting oxidant concentration (LOC). Previously the LOC was also called the minimum oxidant concentration (MOC). If a fuel–air mixture contains a concentration of oxygen below the LOC, then a flame will not propagate due to the lack of oxygen, rather than lack of fuel. LOCs of numerous gases (with N_2 or CO_2 diluent) can be found in NFPA 69 (1997b). If a value for an LOC cannot be found in the literature from experimental determination, it can be estimated from the LFL (with nitrogen as diluent) and the stoichiometry for a reaction by calculating the number of moles of oxygen needed to completely react with one mole of fuel (Bodurtha 1980, Crowl and Louvar 1990):

$$\text{LOC} = \text{LFL} \times (\text{moles } O_2/\text{moles fuel}) \tag{4-2}$$

Autoignition Temperature (AIT)

The autoignition temperature of a gas or vapor is the minimum temperature at which it will ignite in air without any external source of ignition (e.g., a flame, spark, hot surface etc.). It is often also called the spontaneous-ignition or self-ignition temperature. The measurement of AIT depends on many factors, such as molecular structure of the gas/vapor, combustible/oxidant ratio, volume of vessel or surface/volume ratio, vessel surface conditions, pressure of the system (including injection pressure during an AIT test), presence of catalytic material, ignition delay, and flow conditions. Thus, a specified AIT should be interpreted as applying only to the experimental conditions employed in its determination. In conducting tests to determine AITs for a process, it is essential to use experimental conditions as close as possible to process conditions.

Composition affects the AIT. Mixtures that are excessively fuel-rich or fuel-lean generally ignite at higher temperatures than those at intermediate compositions. In an homologous series of organic compounds, the AIT decreases with increasing molecular weight. Also, larger experimental apparatus volumes decrease the AITs; an increase in pressure decreases AITs; and increases in oxygen concentration decrease the AITs. A discussion of the factors affecting AITs is presented by Bodurtha (1980). AIT data for many chemicals can be found in Table 5-1 and the NFPA Fire Protection *Guide to Hazardous Materials* (NFPA 1997a).

Gases and vapors with low AITs may reignite on the protected side of a flame arrester if they are not sufficiently cooled. See Section 5.3.1 for a discussion of the effect of AIT on DDA performance.

Minimum Ignition Energy (MIE)

Initiation of flame propagation in a combustible mixture at a temperature below the AIT requires an ignition source of adequate energy and duration to overcome radiative and conductive heat losses to the cooler surrounding atmosphere. This is called the minimum ignition energy (MIE), and the lowest value is found at a certain optimum mixture (usually, but not always, near the stoichiometric concentration). Values of MIE for many gases are listed in NFPA 497 (1997c).

Minimum Igniting Current (MIC) Ratio

The minimum igniting current (MIC) ratio is the ratio of the minimum current required from an inductive spark discharge to ignite the most easily ignitable mixture of a gas or a vapor, divided by the minimum current required for an inductive spark discharge to ignite methane under the same test conditions. The MIC ratio is used, as an alternate to the maximum experimental safe gap (MESG), to classify flammable liquids, gases, or vapors for locations which contain electrical installations in chemical process areas, as presented in NFPA 497 (1997). For Class I, Group B chemicals the MIC ratio is less than or equal to 0.40; for Group C chemicals the MIC ratio is greater than 0.40 and less than or equal to 0.80; and for Group D chemicals the MIC ratio is greater than 0.80. If the European zone method is used, for Group IIC chemicals the MIC ratio is less than or equal to 0.45; for Group IIB chemicals the MIC ratio is greater than 0.45 and less than or equal to 0.80; and for Group IIA chemicals the MIC ratio is greater than 0.80.

Sources of Ignition

As mentioned previously, in order to initiate flame propagation, an ignition source of adequate energy and duration is required. Ignition sources can include:

1. *Thermal sources:* flames, hot surfaces, autoignition, compression, internal combustion engines, plasmas
2. *Electrical sources:* static electricity, electrical current, lightning, stray currents (radiofrequency electromagnetic radiation, overhead high voltage transmission lines, galvanic and cathodic protection stray currents)
3. *Mechanical sources:* friction, impact, vibration, metal fracture
4. *Chemical reaction sources:* catalysis, reaction with powerful oxidants, reaction of metals with halocarbons, thermite reaction, thermally unstable materials, accumulation of unstable materials, pyrophoric materials, polymerization, decomposition, heat of adsorption, water reactive solids, incompatible materials.

Discussions of the above sources of ignition and others can be found in the books by Bond (1991), and *Guidelines for Engineering Design for Process Safety* (CCPS 1993).

The location and energy level of an ignition source has a profound effect on flame propagation and the ability of a DDA to quench the flame.

4.1.3. Decomposition Flames

Decomposition reactions are a special class of propagating flames where a molecule can undergo spontaneous exothermic reaction. The most widely reported incidents in which decomposition reactions occur are for acetylene where decomposition primarily gives carbon and hydrogen, as shown in the following reaction:

$$C_2H_2 \rightarrow 2C + H_2 + \text{Heat} \qquad (4\text{-}3)$$

This presents a significant risk in processes involving acetylene as decomposition can occur at pressures as low as atmospheric.

A number of other gases can undergo reactions that produce decomposition flames—for instance, ethylene, ethylene oxide, methyl nitrate, ethyl nitrate, and hydrazine (CCPS 1993).

4.2. Dynamics of Flame Propagation

4.2.1. Burning Velocity and Flame Speed

Fundamental, laminar, and turbulent burning velocities describe three modes of flame propagation (see the Glossary for definitions). The fundamental burning velocity, S_u, is as its name implies, a fundamental property of a flammable mixture, and is a measure of how fast reactants are consumed and transformed into products of combustion. Fundamental burning velocity data for selected gases and vapors are listed in Appendix C of NFPA 68 (1998).

Laminar flames propagate at a velocity determined by the chemical and thermodynamic properties of the flammable mixture and the fundamental burning velocity, S_u. A schematic of the regions associated with a laminar flame is shown in Figure 4-1. As the flame front propagates through the unburned mixture, it is first preheated by energy transferred from the exothermic flame region. Once exothermic reaction starts, the temperature increases rapidly and the gas density decreases. This expansion of the burned products is accompanied by an increase in the gas velocity. The laminar burning velocity, S_L, is influenced by both thermal diffusion and the mass diffusion of radicals (OH, H, O, CO, etc.) toward the unburned gas. Stull (1977) states that the laminar burning velocities of hydrocarbon–air mixtures increase from the lower limit to a maximum on

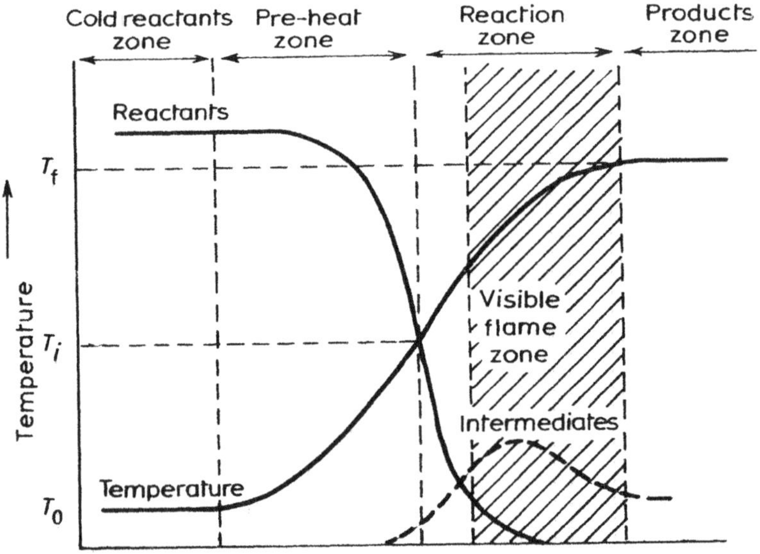

FIGURE 4-1. Concentration and temperature profiles through a premixed flame.

the fuel-rich side of stoichiometric concentration, then decline to a low value at the upper flammable limit. Stoichiometric hydrocarbon–air mixtures show a steady increase in the laminar burning velocity as the initial temperature of the mixture increases, and the laminar burning velocity decreases as the pressure increases.

Once gas velocities exceed a critical value, significant and random fluctuations in the local flow velocity with time are observed to develop, and the velocity is termed a turbulent burning velocity, S_t. It is still possible to identify a mean value for the flow velocity, but there will also now be a random component. The motion of unburned gases can increase the local turbulence intensity by wall boundary layer effects. A positive feedback mechanism is established in which a turbulent flame continually accelerates as it burns into the flow field, which is modified as a consequence of combustion at an earlier time. In turbulent combustion, a laminar flame front becomes distorted (wrinkled) because of the complexity of the flow field. If locally the mixture still propagates with the laminar burning velocity, S_L, the mass burning rate will increase by the ratio of the surface area

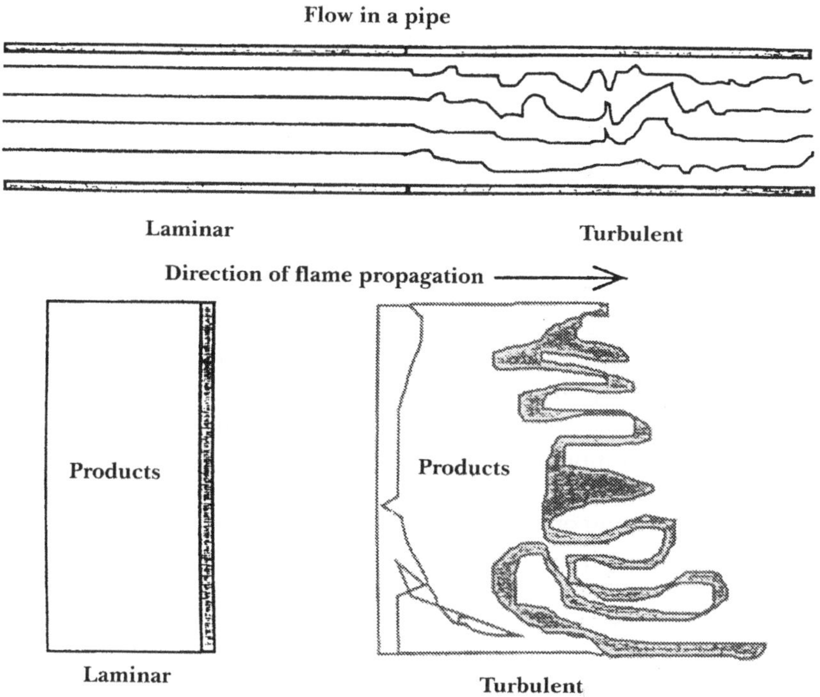

FIGURE 4-2. Sketch of differences in the local direction (upper) and flame front topography (lower) between a laminar and turbulent flame.

increase, as shown in Figure 4-2. The result is equivalent to a faster turbulent burning velocity, S_t, whose magnitude is related to the local turbulence intensity as well as other turbulence parameters.

The consequence is that the rate of production of volume of burned products is greater due to the density decrease resulting from the reaction. As the products expand this causes the unburned mixture to move as well. The flame is then seen to move forward with a higher apparent velocity, V_f, the sum of the mean unburned gas velocity, u_g, and the turbulent burning velocity, S_t. V_f is called the flame velocity (flame speed).

Laminar burning can become turbulent burning if there is sufficient run-up length in the piping and also if there are turbulence-producing elements in the piping such as elbows, tees, valves, etc.

Flashback into a pipe from which a flammable mixture discharges can occur if the gas velocity near the wall falls below the burning velocity (Bodurtha 1980, Lewis and von Elbe 1987). The diameter of the pipe must be larger than the quenching (critical) diameter for a flashback to occur (Bodurtha 1980). Usually, quenching diameters are much smaller than industrial vent pipe diameters. In turbulent flow, the conditions near the center of a vent stream may lead to flashback there as well (Grumer 1958). Most experimental work on flashback has been done with laminar flow compared to turbulent flow conditions. Flashback flame speeds are highest at about the stoichiometric concentration in air. Because mixtures within the flammable range can be ignited accidentally by an external ignition source, they should be vented such that their discharge velocity prevents flame propagation back into the plant equipment.

Flame speed also has an effect on whether a flame will stabilize on a flame arrester element and continue to burn, possibly resulting in the destruction of the arrester element. A stabilized flame occurs on an arrester element if the gas velocity is equal to or just slightly higher than the speed of a deflagration, and there is sufficient time in this condition to allow the flame to travel from the ignition source to the flame arrester (Lapp and Werneburg 1999). Therefore, flame stabilization is dependent on the necessary combination of flow, gas/vapor–air mixture, and time. Flame flashback speed values can range from very high (1800 m/s) to very low (1 m/s), depending on various factors such as pipe configuration, operating pressure, gas/vapor–air mixture ratio, and others. The majority of flashbacks in any system are extinguished instantly against the arrester element because the burning rate is higher than the mixture (fuel) supply rate. Various experimental methods are used for testing stabilized endurance burning, and these are discussed further in Chapter 8.

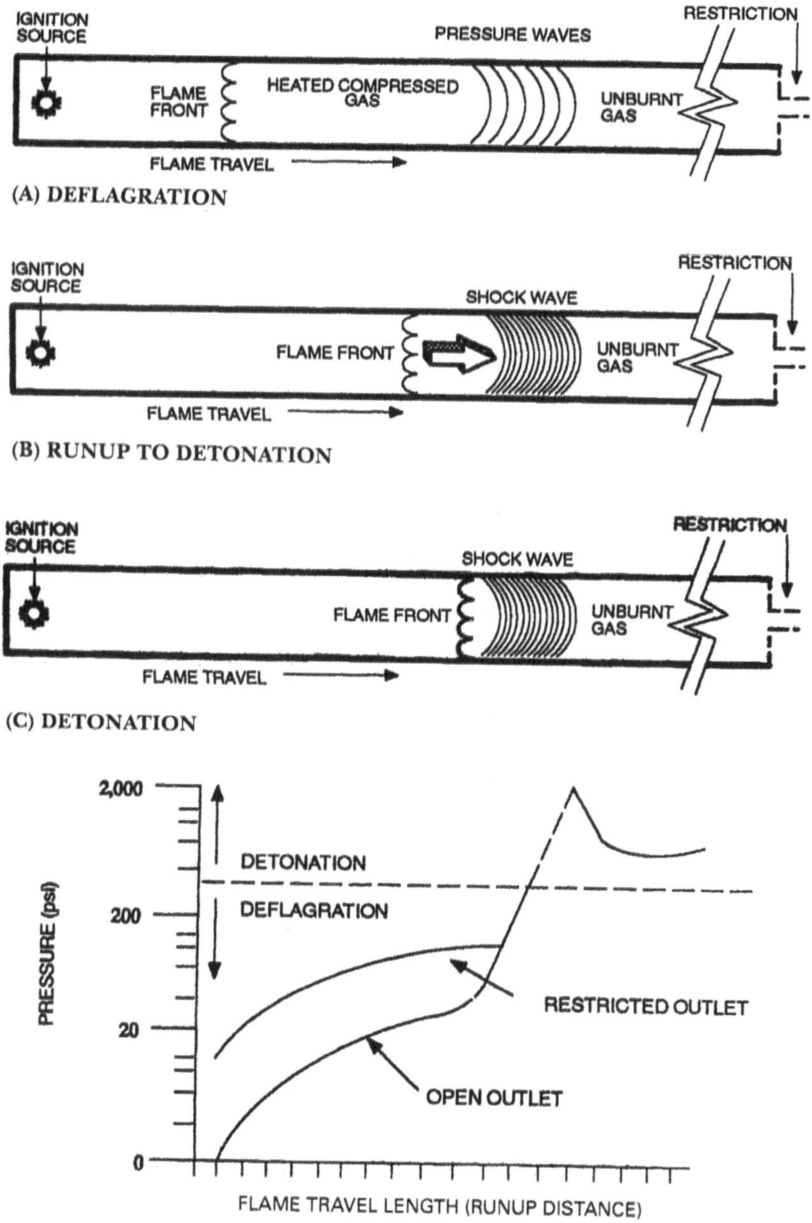

FIGURE 4-3. Differences between deflagration and detonation flame fronts.

4.2.2. Flame Acceleration and Deflagration-to-Detonation Transition (DDT)

Pipeline deflagrations and detonations can be initiated by various ignition sources. The flame proceeds from a slow flame through a faster accelerating turbulent flame to a point where a shock wave forms and a detonation transition occurs, resulting in an overdriven detonation (see Figure 4-3). A stable (steady state) detonation follows after the peak overdriven detonation pressure subsides.

Apart from the direct initiation of detonation by a point source, such as an electric spark or explosive, two main scenarios have been examined to date. These are the natural acceleration of a flame in a long pipe, and the forced acceleration of a less confined mixture within a region containing obstacles. In both cases, gas phase turbulence enhances combustion rates to a point where a shock wave is formed ahead of the flame front. If further flame acceleration occurs, the leading shock wave is strengthened until a transition to detonation occurs. In essence, deflagration-to-detonation transition (DDT) is the transient phenomenon resulting from the acceleration of a deflagration flame to detonation via combustion-generated turbulent flow and compressive heat effects. At the time of transition, a volume of precompressed, turbulent gas ahead of the flame front detonates, developing unusually high velocity and overpressure. The overpressure depends on the degree of precompression due to deflagration before transition occurs, and can be enhanced by shock wave reflections. During the DDT, the initial peak pressure reached is higher than the final pressure reached when the stable detonation phase occurs, and the detonation wave is described as "overdriven." The detailed mechanisms by which DDT occurs is still the subject of debate, as discussed, for example, by Lee (1977) and Sichel (1992).

Deflagration-to-detonation transition can be induced by shock waves. In a detonation, the chemical reactions are initiated in a different way from the corresponding reaction in a flame. Flame combustion is strongly dependent on heat diffusion ahead of the energy release zone. This process is, of course, much more complicated in a turbulent flame. In detonations, however, the reactions are initiated by the pressures and temperatures associated with the shock wave (i.e., the von Neumann peak and related induction zone). If a sufficiently strong shock front is formed ahead of an accelerating flame, the shock-induced reactions (often called autoignition) can lead to the formation of a so-called "hot spot" (Lee 1977, Sichel 1992, and Khoklov et al. 1999), and if the conditions are right, this leads to a second shock wave that rapidly manifests itself as an overdriven detonation.

Thibault et al. (2000) present test data on deflagration and detonation of ethylene oxide (EO) vapor in pipelines. The tests were conducted in two parts with one set of tests in 5-cm and 30-cm pipes with low energy ignition, and in the second set of tests in 30-cm pipe with forced detonation. In a series of 30-cm pipe tests employing low-energy ignition and otherwise increasingly severe conditions, a DDT occurred, partially destroying the test equipment. Analysis of the data suggests that the DDT occurred near the end of the pipe, and based on the pressures inside the pipe just prior to the DDT, it is estimated that the maximum pressure at the end of the pipe was approximately 643 bars absolute (approximately 9300 psia). Since the initial pressure in the pipe was 4.7 bara (68 psia), these data show that unusually high pressures can be generated by a DDT.

Chatrathi et al. (2001) recently reported some experiments on flame propagation in industrial scale piping. They presented data on deflagration propagation in three sizes of pipes (6-inches, 10-inches, and 16-inches) and three fuels (propane, ethylene, and hydrogen). The effects of bends were evaluated, but other piping system components were not evaluated. The conclusions from this work are as follows:

1. Flame speed and pressure generation in a pipeline deflagration are strongly dependent on fuel composition. As lower and upper flammability limits are approached, the flame can still propagate slowly, but without the generation of significant pressure. This latter point is of significance when pressure is employed as the indicator of a deflagration or detonation.
2. Pipe diameter has an effect on flame propagation. It is minimal in the range of *L/D* from 1 to ~50. In this section of the pipe, the flame velocity is not affected by the diameter. Beyond an *L/D* of 50, flame speed increases with pipe diameter.
3. Flame speed is proportional to the fundamental burning velocity; that is, as the fundamental burning velocity increases, the flame speed increases.
4. Bends in the pipeline have some effect on flame propagation behavior, but it is less than anticipated. With normal hydrocarbons (propane), a 90-degree bend causes a temporary rise in flame speed that does not persist. A bend at a high *L/D* (further from the point of ignition), however, does result in significantly higher pressures. Pressure development due to bends reaches levels as high as 150 barg (approximately 2200 psig).
5. The length-to-diameter ratios at DDT were found to be as expected. They ranged from 60 to 70, decreasing with increasing velocity and slightly decreasing with pipe diameter.

4.2.3. Detonations

Detonation is propagation of a combustion-driven shock wave at a velocity at or above the speed of sound in the unreacted medium as measured at the flame front (see Figure 4-3). The wave is sustained by chemical energy released by shock compression and ignition of the unreacted medium (fuel). The flame front is coupled in time and space with the shock front, and there is no pressure increase significantly ahead of the shock-flame front. Propagation velocities in the range of 1000–3500 m/s may be observed depending on the gas/vapor–air mixture, initial temperature and pressure, and type of detonation. Detonations may be initiated within limits of fuel concentration known as the "detonable limits" (which are narrower than flammable limits), either directly using a detonation initiator or via DDT. The following detonation phenomena have been observed:

Galloping detonation is a detonation that periodically fails and reinitiates during propagation. This type of detonation is typically observed in near-limit mixtures (they have been observed near the lean, and possibly near the rich limit). Since it reinitiates via DDT, a galloping detonation is periodically overdriven and results in large overpressures at periodic distances along a pipe. Over these periodic cycles the wave oscillates between a fast deflagration and a leading shock, transition to overdriven detonation, and a short-lived apparently steady detonation phase. There are instances where this has lead to repeated wall failure at regular intervals along the length of pipes containing a galloping detonation.

Overdriven detonation is the condition that exists during a DDT before a state of stable detonation is reached. Transition occurs over the length of a few pipe diameters and propagation velocities up to 2000 m/s have been measured for hydrocarbons in air. This is greater than the speed of sound as measured at the flame front. Overdriven detonations are typically accompanied by side-on pressure ratios (at the pipe wall) in the range of 50–100. A severe test for detonation flame arresters is to adjust the run-up distance so that DDT occurs at the arrester, subjecting it to the overdriven detonation impulse.

Besides being generated in piping, overdriven detonations can also be generated at the outer walls of bends in pipes. The duration of the maximum overpressure is, however, usually relatively short. Overdriven detonations are usually perceived as the worst case event that can occur during a pipeline detonation. This is correct as far as maximum overpressure is concerned. However, the short duration may limit the total impulse and mitigate the consequences. During flame arrester tests, some restricted outlet flame tests may prove more taxing on the arrester performance with the arrester mitigating overdriven detonations, but passing lower (deflagration) flames.

Spin detonation is a detonation that propagates with a spiral trajectory along a pipe. This type of detonation may be observed over a relatively wide range of mixture compositions and pipe diameters at which the detonation cell (a characteristic property of gas–air mixtures) is larger than the pipe diameter. The trajectory results in a locally overdriven detonation front that constantly generates large overpressures and temperatures along a pipe wall.

Stable detonation is a fully developed detonation wave propagating at a constant velocity equal to the speed of sound. For hydrocarbons in air this velocity is typically in the range of 1600–1900 m/s. Typical side-on pressure ratios (as seen at the pipe wall) are in the range of 18–30, while face-on pressure ratios normal to the direction of propagation may greatly exceed 100 due to momentum flow. The characteristics of stable detonations are predictable using the Chapman–Jouguet (CJ) theory as described in books on detonation (Nettleton 1987). Stable detonations that have propagated over large distances are referred to as "long run" stable detonations. They impart a large integrated flow of heat and momentum to the flame arrester.

Detonation Theory

A detonation shock wave is an abrupt gas dynamic discontinuity across which properties such as gas pressure, density, temperature, and local flow velocities change discontinuously. Shock waves are always characterized by the observation that the wave travels with a velocity that is faster than the local speed of sound in the undisturbed mixture ahead of the wave front. The ratio of the wave velocity to the speed of sound is called the Mach number.

The Chapman–Jouguet (CJ) theory is a one-dimensional model that treats the detonation shock wave as a discontinuity with infinite reaction rate. The conservation equations for mass, momentum, and energy across the one-dimensional wave gives a unique solution for the detonation velocity (CJ velocity) and the state of combustion products immediately behind the detonation wave. Based on the CJ theory it is possible to calculate detonation velocity, detonation pressure, etc. if the gas mixture composition is known. The CJ theory does not require any information about the chemical reaction rate (i.e., chemical kinetics).

An improvement on the CJ model is the ZND (Zeldovich, von Neumann, and Doring) model, which takes the reaction rate into account (Nettleton 1987, Glassman 1996, Lewis and von Elbe 1987). The ZND model describes the detonation wave as a shock wave, immediately fol-

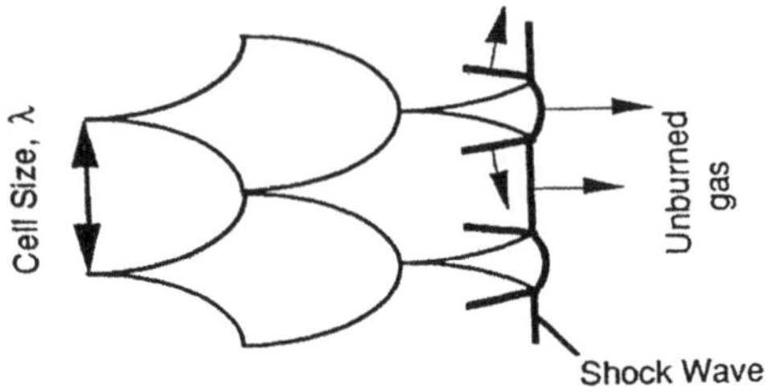

FIGURE 4-4. Pattern of a detonation cell.

lowed by a reaction zone (i.e., flame). The thickness of this zone is given by the reaction rate. The ZND model gives the same detonation velocities and pressures as the CJ model; the only difference between the two models is the thickness of the wave.

Two different characteristic length scales are currently being used to assess the relative detonability of flammable gases. One length scale is the chemical induction length, defined loosely as the chemical reaction time multiplied by the velocity of the detonation wave. Chemical induction lengths, which have orders of magnitude ranging from 0.01 to 1 cm for stoichiometric hydrocarbon–air mixtures, are determined either by shock tube measurement or by chemical kinetic calculations.

The other detonability length scale is the detonation cell width, λ (also called cell size) which is the transverse dimension of diamond shaped cells generated by the transverse wave structure at a detonation front. It has a "fish scale" pattern (see Figure 4-4). Detonation cell widths are usually measured by the traces (soot) deposited on smoke foils inserted in test vessels or piping surfaces. The more reactive the gas–air mixture, the smaller is the cell size. The same is true for chemical induction length as a qualitative measure of detonability. The cell width, λ, is a parameter that is of practical importance. The transition from deflagration to detonation, propagation, and transmission of a detonation, can to some extent be eval-

TABLE 4-3
Detonation Cell Widths of Some Gases

Gas/Vapor	Cell Width (cm)	Gas/Vapor	Cell Width (cm)
Acetylene	0.98	Propylene	5.4
Hydrogen	1.5	Propane	6.9
Ethylene	2.8	Hydrogen Sulfide	10.0
n-Butane	5.8 to 6.2	Methane	28.0
Ethane	5.4 to 6.2		

uated based on the knowledge of the cell width of the mixture. It should be pointed out, however, that the cell width as a measure of detonability is not an exact number. In the literature a variation of a factor of two is often found. When using cell widths for estimation of limiting conditions for successful propagation or transmission, they should be regarded as approximate values. Hence, safety factors should be used. Table 4-3 lists detonation cell widths for a number of stoichiometric gas–air mixtures (SFPE 1988). Lees (1996) presents a number of equations for calculating the induction length and the cell width.

Detonation Limits

Detonation is to some extent unique among propagating combustion phenomena in that thermodynamic predictions of detonation properties are very accurate. Despite the success of the Chapman–Jouguet theory in predicting the parameters of an established detonation, such as velocity and pressure, the theory is unable to give any guidance on whether detonation can be initiated readily in a mixture for a given geometry and other initial conditions (e.g., pressure and temperature) in a pipe of given diameter. The cause of this lack of understanding is the complex interaction between the reaction chemistry and the gas flow dynamics, as characterized by the transverse wave structure.

Initiation of a steady-state detonation requires that the self-sustaining multidimensional structure must develop. In a related way, if transverse structure can be eliminated, then the detonation fails. Thus a detonation cannot propagate in a pipe whose internal diameter is close to the detonation cell width. Test results by Dupre et al. (1985, 1990) with a range of pipe diameters suggest that that detonation cannot propagate along a circular pipe when the cell width is greater than the circumference of the pipe. Also, the cell length can be equated very crudely to some 200 times the induction zone length in the von Neumann or induction zone, and the cell width is some 0.6 times the cell length.

Thus it is possible to estimate order of magnitude limits for detonation propagation using calculated CJ induction zone lengths or measured cell size data. These were limits for established detonations propagating into pipes of decreasing diameter. Variations in the detonability of different mixtures in different pipe geometries are thus intimately linked to the initial chemical and physical properties of the mixture.

Despite the number of existing studies on detonation over the past decades, it is unfortunate that, apart from the most common hydrocarbon fuels and hydrogen, the ability to predict detonation limits is still virtually nonexistent. Nettleton (1987) does present some equations for estimating

TABLE 4-4
Detonation Limits (Vol %) for Confined and Unconfined Explosions and Flammability Limits (Vol %) in Oxygen and Air

Fuel	*Confined detonation limit, O_2*		*Confined detonation limit, air*		*Unconfined detonation limit, O_2*		*Unconfined detonation limit, air*		*Flammability limit, air*		*Flammability limit, O_2*	
	Lower	*Upper*	*Lower*	*Upper*	*Lower*	*Upper*	*Lower*	*Upper*	*Lower*	*Upper*	*Lower*	*Upper*
C_2H_6	3.60	46.4	2.87	12.20	11.0	39.0	4.0	9.2	3.0	12.4	3.0	66.0
C_3H_8	2.50	42.5	2.57	7.37	7.0	31.0	3.0	7.0	2.1	9.5		
nC_4H_{10}	2.05	38.0	1.98	6.18			2.5	5.2	1.8	8.4		
nC_8H_{18}	1.55	17.3	1.45	2.85					0.95			
C_2H_4	4.10	60.0	3.32	14.70	9.2	51.0			2.7	36.0	2.9	80.0
C_3H_6	2.50	50.0	3.55	10.40	6.7	37.0	3.5	8.5	2.4	11.0	2.1	53.0
C_2H_2	2.90	88.8	4.20	50.0	6.7	68.0			2.5	80.0		
$CH_3C{\equiv}CH$					5.7	49.0			1.7			
MAPP (37.8% $CH_3C{\equiv}CH$)							2.9	10.1				
Petrol†			~5.6	~9.4								
CH_3OH	9.50	64.5							6.7	36.0		
C_2H_5OH			5.1	9.8					3.3	19.0		
$C_2H_5OC_2H_5$	2.6	>40	2.8	4.5	4.7	29.0			1.9	36.0	2.0	82.0
Cyclo C_3H_6									2.4	10.4	2.5	60.0
Cyclo C_6H_{12}	1.4	29.0			6.7	39.0			0.57	7.8		
C_6H_6	1.55	36.0	1.60	5.55					1.3	7.9		
Xylene	1.05	26.5							1.1	6.4		
CH_3COCH_3	3.3	40.0							2.6	13.0		
CH_3CHO					13.0	48.0			4.0	60.0		
H_2	15.0	90.0	18.3	58.9					4.0	75.0	4.0	95.0

Source: Nettleton, M.A. 1987. *Gaseous Detonations: Their Nature, Effects, and Control.* Chapman and Hall, New York, NY.

the lower and upper detonation limits in a confined situation for a number of hydrocarbon fuels mixed with both air and oxygen.

Experimental data are available for detonation limits for a limited number of fuel–air and fuel–oxygen mixtures at atmospheric pressure in both confined and unconfined situations. These are presented in Table 4-4 (Nettleton 1987).

Relatively little information is available on the influence of initial pressure and initial temperature on detonation limits, in confined situations. The general effect of an increase in initial pressure is to widen the detonations limits, and the same applies to an increase in the initial temperature. The widening is more marked for the upper limit.

For a more detailed discussion of detonations and detonation limits, the books by Nettleton (1987), Glassman (1996), and Lewis and von Elbe (1987) are recommended.

4.3. Ignition and Quenching

For ignition of a flammable mixture to occur, several factors are necessary:

- An ignition source must have sufficient energy and duration to initiate a reaction
- The rate of heat generation by reaction must exceed the rate of heat loss to allow the reaction to propagate
- The concentration of the gas/vapor must be in the flammable range
- The ignition source energy must be above the minimum ignition energy (MIE) of the gas or vapor

Two properties of gases and vapors that may determine when an ignition can occur are the minimum ignition energy (MIE) and the autoignition temperature (AIT). These are discussed in Section 4.1.2 above. The MIE is a function of the pressure, temperature, and composition of a fuel–oxidant mixture.

There is a relationship between the MIE of a gas or vapor and the quenching distance, as shown in Figure 4-5 (van Dolah and Burgess 1974). These data are for a large variety of chemicals with oxygen varying between 21 and 100 volume percent and pressure between 0.1 to 2 atmospheres.

Quenching distance, Q_d, is the dimension of a flow channel that just prevents passage of a flame. It is generally determined experimentally in the following manner. A premixed flame is established on a burner port and the gas flow is suddenly stopped. If the flame propagates down the tube into the gas supply source, a smaller tube is then tried. The tube diameter is progressively decreased until the flame cannot propagate back into the source. This is then the quenching distance. If the flow channel is

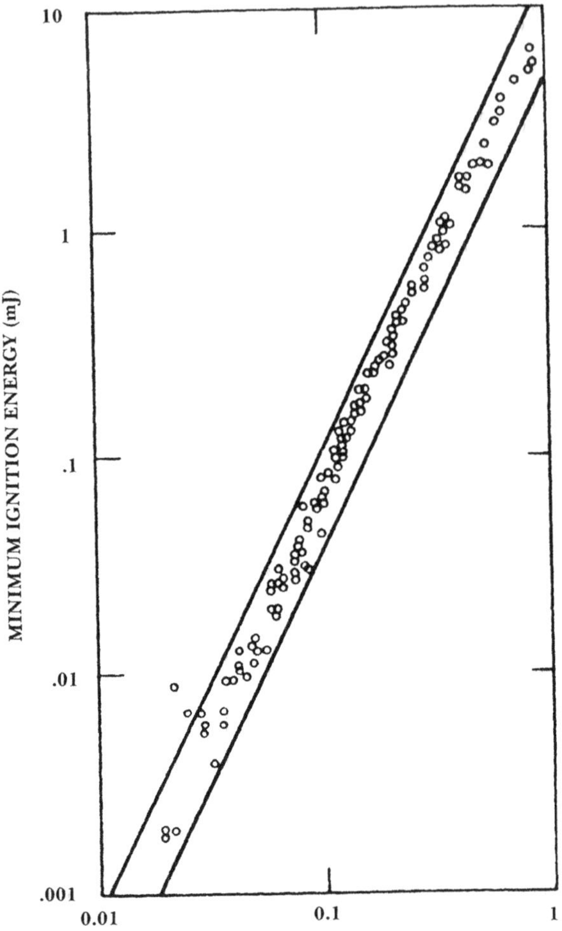

FIGURE 4-5. The relationship between minimum ignition energy (MIE) and quenching distance

noncircular, it is called the quenching distance. If the channel is circular, it is called the quenching or critical diameter, Q_{cr}. In rectangular burner tubes, the quenching distance—that is, the distance within which two plates must be brought to prevent flashback—is smaller than the quenching (critical) diameter in cylindrical tubes.

It has been found experimentally that increasing the temperature reduces the quenching distance. However, sufficient data are not available to develop any specific correlation (Glassman 1996).

It has been established that the quenching distance increases as pressure decreases as shown in the following equation (Glassman 1996):

$$Q_d \sim 1/P \tag{4-4}$$

In fact, this correlation is almost exact for many compounds. For some combustible gases and vapors, P sometimes has an exponent somewhat less than 1.

4.4. Theoretical Basis for Flame Arrester Design and Operation

Some attempts have been made to provide a theoretical basis for the design of flame arresters (Wilson and Atallah 1975, Wilson and Flessner 1978). Despite these works and the significant effort expended on flame arrester development, there is still no rigorous quantitative theoretical basis for their design and operation. Qualitatively, they must operate by reducing the momentum associated with the gas motion in a deflagration or detonation, while reducing the temperature of the products of combustion to the point where reaction in the unburned mixture can no longer be self-sustaining, and the flame is quenched.

Any design guidelines that do exist are usually based on empirical observations and correlation rather than fundamental theory. As an example, it has been experimentally shown that the use of narrow channels in crimped metal flame arresters provide significant resistance to gas flow together with a large heat transfer surface area to provide cooling. Therefore, it is very important to test any new flame arrester application using any of the approved testing protocols (USCG, UL, FM, CEN, or Canadian Z343 standards) to verify if the arrester will perform satisfactorily.

4.5 References

Baker, W. E. et al. 1983. *Explosion Hazards and Evaluation.* Elsevier, New York, NY.

Bodurtha, F. T. 1980. *Industrial Explosion Prevention and Protection.* McGraw-Hill, New York, NY

Bond, J. 1991. *Sources of Ignition–Flammability Characteristics of Chemicals and Products.* Butterworth-Heinemann, Boston, MA.

Bradley, J. N. 1969. *Flame and Combustion Phenomena.* Methuen, London, England, UK.

Britton, L. G. 1996. Operating Atmospheric Vent Collection Headers Using Methane Gas Enrichment. *Process Safety Progress,* 15(4), 194–212.

CCPS. 1993. *Guidelines for Engineering Design for Process Safety.* Center for Chemical Process Safety, American Institute of Chemical Engineers, New York, NY.

Chatrathi, K., Going, J. E., and Grandestaff, B. 2001. Flame Propagation in Industrial Scale Piping. Proc. AIChE 35th Annual Loss Prevention Symposium, Paper 1D, April 22–26, 2001, Houston, TX.

Coward, H. F. and Jones, G. W. 1952. *Limits of Flammability of Gases and Vapors.* Bureau of Mines Bulletin 503, US Department of the Interior, Washington, DC.

Crowl, D. A. and Louvar, J. F. 1990. *Chemical Process Safety: Fundamentals with Applications.* Prentice-Hall, New York, NY.

Dupre, G., Knustautus, R., and Lee, J. H. 1985. *AIAA Prog. Astronaut. Aeronaut.,* 106, 244–259.

Dupre, G., Peraldi, O., Joannon, J., Lee, J. H., and Knustautus, R., 1990. *AIAA Prog. Astronaut. Aeronaut.,* 133, 156–169.

Glassman, I. 1996. *Combustion.* 3rd ed. Academic Press, New York, NY.

Green, D. W., and Maloney, J. O. 1997. *Perry's Chemical Engineers' Handbook.* 7th edition. McGraw-Hill, New York, NY.

Griffiths, J. F., and Barnard, J. A. 1995. *Flame and Combustion.* 3rd ed. Blackie Academic & Professional/Chapman & Hall, New York, NY.

Grumer, J. 1958. Flashback and Blowoff Limits of Unpiloted Turbulent Flames. *Jet Propulsion,* 28(11), 756–758 (November 1958).

Khoklov, A. M., Oren, E. S., and Thomas, G. O. 1999. *Combustion and Flame,* 117, 323–339.

Lapp, K. and Werneburg, H. 1999. Detonation Flame Arrester Qualifying Application parameters for explosion Protection in Vapour Handling Systems. *J. Can. Petrol. Technol.,* Special Edition 1949–1999, Paper 94-58.

Lee, J. H. S. 1977. *Ann. Rev. Phys. Chem.,* 38, 75–104.

Lees, F. P. 1996. *Loss Prevention in the Process Industries.* 2nd ed., 2, 17/11-17/12. Butterworth-Heinemann, Boston, MA.

Lewis, B. and von Elbe, G. 1987. *Combustion, Flames, and Explosions of Gases.* 3rd ed. Academic Press, New York, NY.

Mashuga, C. V. and Crowl, D. A. 2000. Derivation of Le Chateler's Mixing Rule for Flammable Limits. *Process Safety Progress,* 19(2), 112–117.

NFPA. 1997a. *Fire Protection Guide to Hazardous Materials.* National Fire Protection Association, Quincy, MA.

NFPA 68. 1998. *Guide for Venting of Deflagrations.* National Fire Protection Association, Quincy, MA.

NFPA 69. 1997b. *Standard on Explosion Protection Systems.* National Fire Protection Association, Quincy, MA.

NFPA 497. 1997c. *Classification of Flammable Liquids, Gases and Vapors and Hazardous (Classified) Locations for Electrical Installations.* National Fire Protection Association, Quincy, MA.

Nettleton, M. A. 1987. *Gaseous Detonations: Their Nature, Effects and Control.* Chapman and Hall, New York, NY.

SFPE. 1988. *The SFPE Handbook of Fire Protection Engineering.* Section 2, Chapter 5, 2-95, National Fire Protection Association, Quincy, MA.

Sichel, M. 1992. Transition to Detonation-Role of Explosion within an Explosion. In *Major Research Topics in Combustion* (Hussaini, Y. A., Kumat, A., and Voight, R. G., eds.), Springer-Verlag, Berlin.

Strahle, W. C. 1993. *An Introduction to Combustion.* Gordon and Breach, New York, NY.

Stull, D. R. 1977. *Fundamentals of Fire and Explosion.* American Institute of Chemical Engineers, New York, NY.

Thibault, P., Britton, L. G., and Zhang, F. 2000. Deflagration and Detonation of Ethylene Oxide Vapors in Pipelines. *Process Safety Progress,* 19(3), 125–139.

Van Dolah, R. W. and Burgess, D. S. 1974. *Explosion Problems in the Chemical Industry.* American Chemical Society, Washington, DC.

Wilson, R. P., Jr. and Attalah, S. 1975. *Design Criteria for Flame Control Devices for Cargo Venting Systems.* US Coast Guard Report No. CG-D-157-75 (August 1975). Available from the NTIS, Springfield, VA 22161.

Wilson, R. P. Jr. and Flessner, M. F. 1979. Design Criteria for Flame Arresters. Paper presented at the American Institute of Chemical Engineers 84th National Meeting, February 28–March 1, 1978, Atlanta, GA.

Zabetakis, M. G. 1965. *Flammable Characteristics of Combustible Gases and Vapors.* Bureau of Mines Bulletin 627. US Department of the Interior, Washington, DC.

5

Deflagration and Detonation Flame Arrester Technology

5.1. Where Flame Arresters May Be Needed

OSHA 1910.106 (based on the 1969 edition of NFPA 30) and NFPA 30 (2000) designate where conservation vents and flame arresters may be needed on storage tanks or process vessels containing flammable or combustible liquids at atmospheric pressure. Sections (b)(2)(iv)(f) and (g) of OSHA 1910.106 state as follows:

> (f) Tanks or process vessels storing Class IA liquids shall be equipped with venting devices which shall be normally closed except when venting to pressure or vacuum conditions. Tanks and pressure vessels storing Class IB and IC liquids shall be equipped with venting devices which shall be normally closed except when venting under pressure or vacuum conditions, or with approved flame arresters. Exemption: Tanks of 3000 bbls. capacity or less containing crude petroleum in crude producing areas; and outside aboveground atmospheric tanks under 1000 gals. capacity containing other than Class IA flammable liquids may have open vents.
>
> (g) Flame arresters or venting devices required in subdivision (f) of this subdivision may be omitted for Class IB and IC where the conditions are such that their use may, in case of obstruction, result in tank damage.

The requirements above and in NFPA 30 must be properly applied after evaluation to ensure that they apply to the tank system concerned. The latest edition of NFPA 30 should be used as it is periodically updated.

NFPA 30 (2000), Section 5.10, applies to vapor recovery (vent manifold) and vapor processing systems where the vapor source operates at pressures from vacuum up to and including 1 psig. Subsection 5.10.7.6 is concerned with flame propagation hazards, but is not specific about installing flame arresters. It states as follows:

> Where there is reasonable potential for ignition of a vapor mix in the flammable range, means shall be provided to stop the propagation of flame through the vapor collection system. The means chosen shall be appropriate for the conditions under which they will be used.

The appropriate protective means can be flame arresters or any of the other protective measured discussed in Chapter 3.

The U.S. Coast Guard regulations 33 CFR Part 154, Subpart E—Vapor Control Systems (1990), originally applied to facilities that collected vapors of crude oil, gasoline blends, or benzene emitted from vessel (ship and barge) cargo tanks. However, the regulations can be, and are being, applied to other chemicals if the facility is approved by the USCG Commandant as meeting the requirements. The USCG regulations are presently being revised to cover other flammable vapors (Schneider 2000).

There are currently no regulations or other legal requirements for installing flame arresters in vapor collection (vent manifold) systems in chemical and petrochemical plants. However, many chemical companies are following the USCG regulations as a guide for other systems where there are no regulatory requirements.

The installation of flame arresters should also be considered for vacuum pumps, activated carbon adsorbers, etc. which emit flammable vapors and/or can serve as ignition sources.

5.2. Types of Flame Arresters

5.2.1. Introduction

This section describes various types of flame arresting elements (matrixes) that are used in fixed element (static) dry type flame arresters, as well as a number of other types. Some of these arresting elements are often used in both deflagration and detonation flame arresters.

5.2.2. Crimped Metal Ribbon

The crimped metal ribbon arresting element, shown in Figure 5-1, is one of the most widely used types, especially for detonation flame arresters. Crimped metal ribbon arresters are made of alternate layers of thin corrugated metal ribbon and a flat metal ribbon of the same width that are wound together on a mandrel to form a cylindrical assembly of many layers of the desired diameter. The thickness of the cylindrical element is equal to the ribbon width. The spaces between the corrugations and the flat ribbon provide multiple small gas passages of approximately triangular shape. Elements can be made in a variety of crimp heights, ribbon metal thicknesses, element thicknesses and diameters. Some of the major advantages of this type of arresting element are: (1) it can be manufactured to within close tolerances, (2) it is sufficiently robust to withstand mechanical and thermal shock, and (3) it has fairly low resistance to flow (pressure drop) because usually only about 20% of the face (cross-sectional area) of the arrester is obstructed by the ribbon. It is important that the layers of ribbon do not spring apart because such movement would increase the crimp height and render the device ineffective. Since the effectiveness in quenching a flame diminishes rapidly with thin arresters, the element thickness should be at least 0.5 inches thick (HSE 1980).

One crimped metal ribbon flame arrester manufacturer has a composite element design consisting of multiple crimped metal ribbon elements with diverter shields (turbulence-inducing devices) between the elements (Enardo n.d.). This design is based on the patent issued to Roussakis and Brooker (1995).

Crimped metal ribbon arresting elements can be made circular, rectangular, or square depending on the shape of the pipe or housing in which they are to be installed. The element is often reinforced by inserting metal rods radially through the assembly.

In the United States and the United Kingdom, crimped metal ribbon arresters may use single or multiple elements with the crimp perpendicular to the ribbon. In Germany, two or three elements (disks) separated by a small gap are used, and the crimp is biased at 45° to the ribbon. The German manufacturer claims that having several shorter height disks make it easier to more completely clean a dirty element. Phillips and Pritchard (1986) indicate that there is no evidence to suggest any advantage for either construction, although the single element with the perpendicular crimp is easier to manufacture. More recent designs include deflectors between element sections to redistribute flow.

A drawback of crimped metal ribbon arrester elements is sensitivity to damage during handling. This should be considered carefully during maintenance of the element. Damage may lead to enlarged channels allowing flame penetration or to channel collapse resulting in increased pressure drop. Therefore, the manufacturer's instructions should be followed strictly during maintenance and cleaning. Another possible problem is that the small channel size may make these arresting elements more sus-

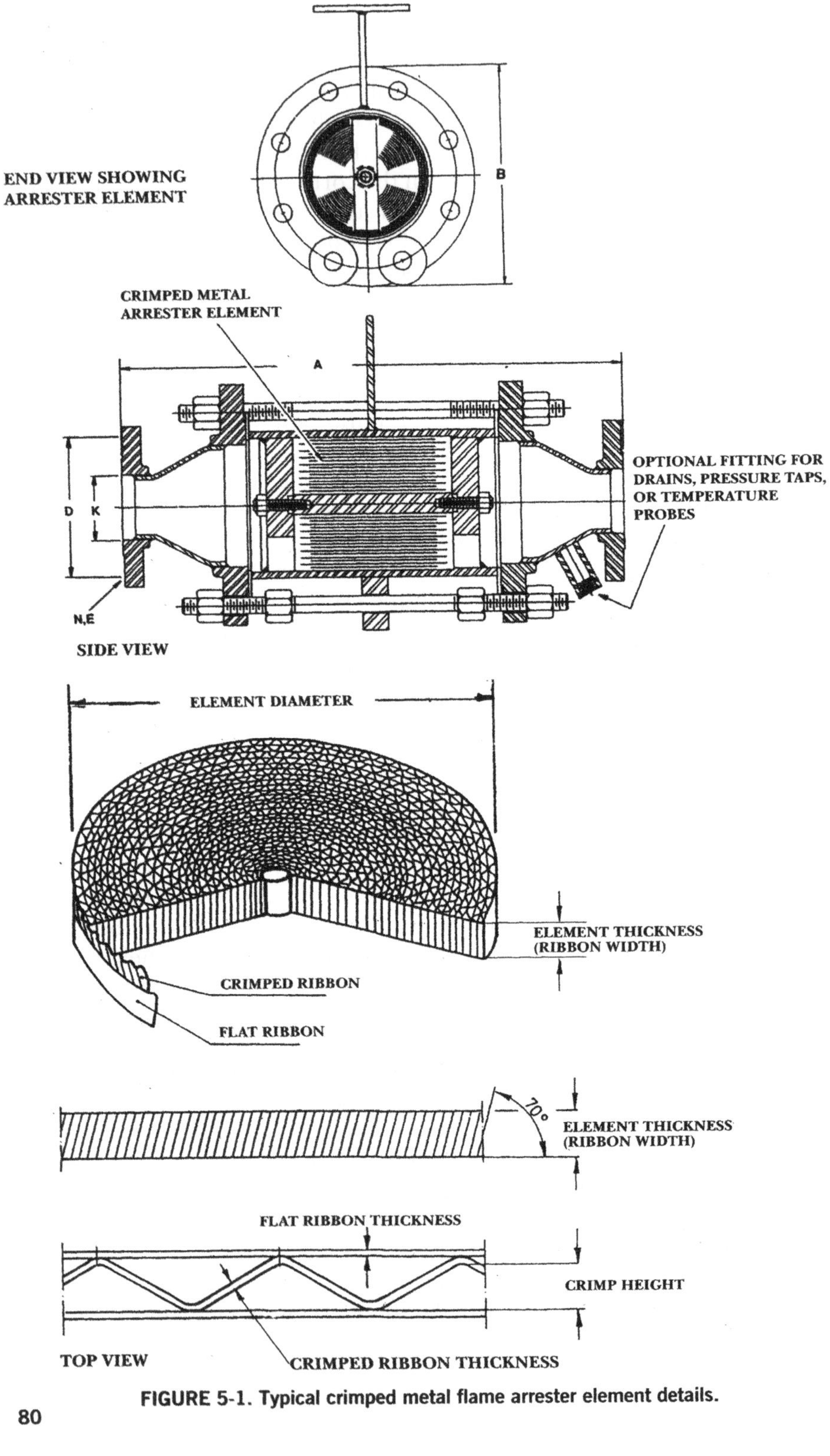

FIGURE 5-1. Typical crimped metal flame arrester element details.

ceptible to fouling due to solids deposition, and regularly scheduled or predictive maintenance is essential when this is a possibility.

Crimped metal ribbon elements are installed in housings in two ways. In the first design the element is removable; it can be removed, cleaned, and reinstalled or replaced without removal of the housing from the piping. In the second design the element is fused into the housing and cannot be removed. In this case, the entire unit (housing and element) must be removed from the piping to clean the element. If the element is damaged, the unit must be replaced. It should be noted that many test protocols (USCG, FM, CEN) will not allow the approval of flame arresters where the element cannot be removed from the housing.

Crimped metal ribbon flame arresters are applicable for both deflagrations and detonations. They are especially used for detonations, since the apertures can be made very small, which is necessary to stop a detonation.

Numerous experimental investigations have been carried out to evaluate the effectiveness of crimped metal ribbon flame arresters to quench deflagrations and detonations for a variety of gases. These are discussed in the following articles and reports: Bjorklund and Ryason (1980), Bjorklund et al. (1982), Broschka et al. (1983), Capp and Seebold (1991), Cubbage (1959), Cubbage (1963), Flessner and Bjorklund (1981), Palmer and Tonkin (1963), Palmer and Rogowski (1968), and Rogowski and Pitt (1976).

5.2.3. Parallel Plate

Parallel plate arrester elements are used in both end-of-line and in-line (vent-line) deflagration arresters. They are not used in detonation arresters, however. These arresters are constructed of unperforated metal plates or rings arranged edgewise to the gas flow and separated from each other by a small spacing. The spacing is maintained by small gaskets or by small nubs that are integral to the plates (Figure 5-2). They are relatively low in cost, robust, and readily dismantled for cleaning. Their main disadvantage is weight, especially in large sizes with housings made of steel or stainless steel (HSE 1980). Large size units may require independent support when mounted on a tank nozzle because of the weight of the unit.

Broschka et al. (1983) report results of experimental tests on parallel plate flame arresters in piping systems. Tests were conducted on 3-inch and 6-inch diameter parallel plate flame arresters installed in 3-inch and 6-inch diameter piping sections using butane–air mixtures to generate a flame. The ignition source was varied from 3 to 43 feet from the flame arrester. The flame speed varied between 0 to 20 ft/s, and when the flame speed was 20 ft/s, the flame passed through the arrester (flame arrester failure).

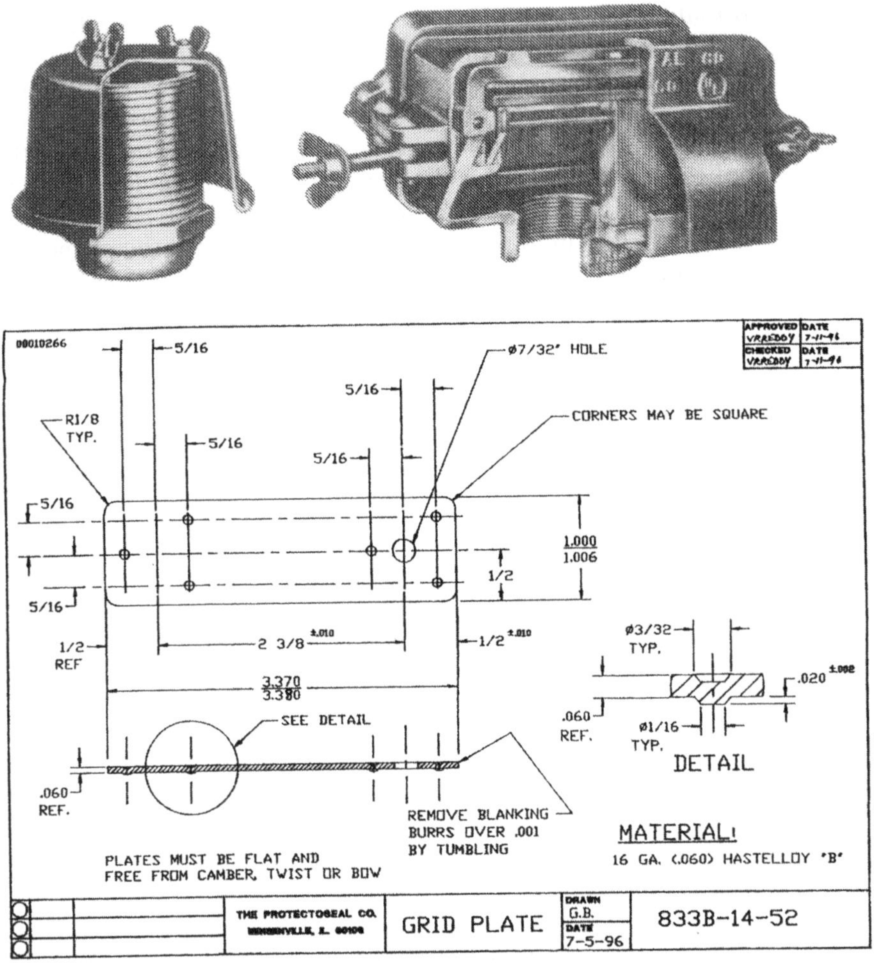

FIGURE 5-2. Parallel plate flame arrester element details. (Source: Protectoseal Company.)

5.2.4. Expanded Metal Cartridge

Expanded metal cartridge elements are composed of a sheet of expanded metal that is wrapped in a fashion similar to a cartridge filter element. Diamond-shaped openings in the expanded metal sheet are not aligned during wrapping so that there is no direct path from one layer to the next. Figure 5-3 shows details of an expanded metal cartridge element. This design tends to reduce the incidence of plugging by suspended solids since these will not be heavily deposited on the inlet face. The elements are nor-

FIGURE 5-3. Expanded metal cartridge flame arrester element details. (Source: Westech Industrial Ltd.)

mally offset, rather than in-line, with respect to the gas flow so that the flow passes radially toward the cartridge axis. This creates a relatively large inlet surface area that further reduces plugging problems. Other advantages include liquid and solids dropout into the external container surrounding the inlet. This feature may make these units suitable for reactive monomer service. Disadvantages include support problems of these units for larger pipe diameters due to their size and weight. Often these must be located at or near grade to facilitate maintenance.

Expanded metal cartridge flame arresters are available for deflagration and detonation applications and are designed for bidirectional flow.

Successful full-scale tests on quenching of deflagrations and detonations using expanded metal cartridge flame arresters were performed to USCG standards on Group C and D gases by Westech Industrial Ltd. (Lapp 1992, Lapp and Vickers 1992).

Expanded metal cartridge elements are manufactured in different configurations. One configuration is that of a cylinder which fits into a housing with offset inlet and outlet connections (Figure 3-7d). The other configuration is that of a "thimble" welded to a flange for insertion in an in-line, "straight-through," housing (Figure 3-7c).

5.2.5. Perforated Plate

Perforated plate arresting elements are used primarily for deflagration flame arresters. The perforated plates are usually metal (stainless steel), but some designs also incorporate perforated refractory disks and gauze pads in combination with metal plates (Zanchetta 1998). The diameters of the holes and the thickness of the plates that are available cover a fairly wide range, but the perforated metal plates most easily obtained for flame arresters have hole diameters and thicknesses similar to coarse gauze flame arresters. Perforated metal arresters have greater mechanical strength and are less likely to overheat than gauze arresting elements, but the proportion of the area of the plate that is available for gas flow is even less than that for the corresponding gauzes (HSE 1980). Figure 5-4a shows details of perforated plate arresting elements.

5.2.6. Wire Gauze

Wire gauzes were used in Sir Humphrey Davy's miners' lamps, and they have been used as flame arresting elements in various applications ever since. These arresters are in the form of single gauzes or a series or pack of gauzes. They are manufactured in a way that ensures that the aperture size is carefully controlled. Single layers of wire gauze have a very limited performance. Gauzes coarser than 28 meshes to the linear inch are ineffective in quenching a flame, and those finer than 60 meshes to the linear inch are liable to become blocked. The main advantages of gauzes are their low cost, ready availability, and the ease of fitting. Their disadvantages include limited effectiveness at quenching high-velocity flames, the ease with which they are damaged, and the resistance of fine gauzes to the flow of gases (high pressure drop).

Gauzes can be combined into multiple-layer packs, and if the gauzes are all of the same mesh width, they are more effective flame arresters than single gauzes; however, the increased effectiveness is limited. Combined packs of a coarse mesh and a fine mesh are less effective flame arresters than the fine gauze alone. A disadvantage of gauze packs is that the good contact required between gauze layers may be difficult to guarantee in practice without fusing and calendering (HSE 1980).

Since gauzes have limited effectiveness in quenching high-velocity flames, they are only used as end-of-line deflagration flame arresters.

Bjorklund et al. (1982) report experimental results on the evaluation of a single 30-mesh gauze screen and a dual 20-mesh gauze screen flame arresters using propane–air and ethylene–air mixtures. The test results are as follows:

1. The single 30-mesh stainless steel flame arrester was effective in arresting flashback flames from all eight fuel–air mixtures tested.
2. The dual 20-mesh stainless steel arrester was effective in arresting flashback from all eight fuel–air mixtures tested except in some ethylene–air tests. It failed in three out of three tests where the flame speed was 4.86 m/s (15.94 ft/s) or greater.

Figure 5-4b shows details of arresting elements of wire gauze and wire gauze packs.

5.2.7. Sintered Metal

Sintered metal is very effective as an arresting element, but it offers a high resistance to gas flow; therefore, it is suitable only for uses where the gas flow is small or high pressure is available (e.g., compressor discharge). Banks of sintered metal flame arresters can be installed in parallel to offset the pressure drop problem. Another disadvantage is that the small apertures have a tendency to block easily, and these flame arresters therefore should be used only with clean gases. One advantage of sintered metal is that it can be produced in a variety of shapes to suit the application. The mounting of sintered metal flame arresters is very critical because the clearance between the arresting element and the housing must be less than the arrester passage dimensions (Howard 1982). If a flame stabilizes on the surface of a sintered metal element, there is a risk that the flame will eventually burn its way through the sintered metal disk. For this reason, these flame arresters may incorporate a pressure- or temperature-activated flow cut-off device (Phillips and Pritchard 1986).

The apertures in sintered metal elements can be made so small that this arrester is able to quench detonations provided that it has sufficient mechanical strength. Particular care is required to ensure a secure anchorage of the sintered element to prevent leakage around the element caused by the impact of the shock wave (HSE 1980).

The main uses of a sintered metal flame arrester are in the sensing heads of flammable gas detectors and in flame arresters for gas welding (oxyacetylene) equipment.

A proprietary sintered metal arrester was made by the Linde Division of Union Carbide Corporation (now Praxair) for use in processes handling acetylene, but is no longer made by Praxair (Dickerman 1999). A sintered metal flashback flame arrester for use on an acetylene cylinder is made by Western Enterprises of Westlake, OH. Figure 5-4c shows a sintered metal flame-arresting element.

5.2.8. Ceramic Balls

Ceramic (alumina) balls are used by one flame arrester manufacturer as the flame-arresting element for detonation arresters (Tornado n.d.). The ceramic balls are contained between stainless grid assemblies. These flame arresters have been tested in accordance with the CSA Z343 standard for NEC Group C and D gases as well as for hydrogen service (see Section 5.3.1 for definition of Groups). They have also been accepted by the U.S. Coast Guard.

Figure 5-4d is a schematic of a ceramic ball flame-arresting element.

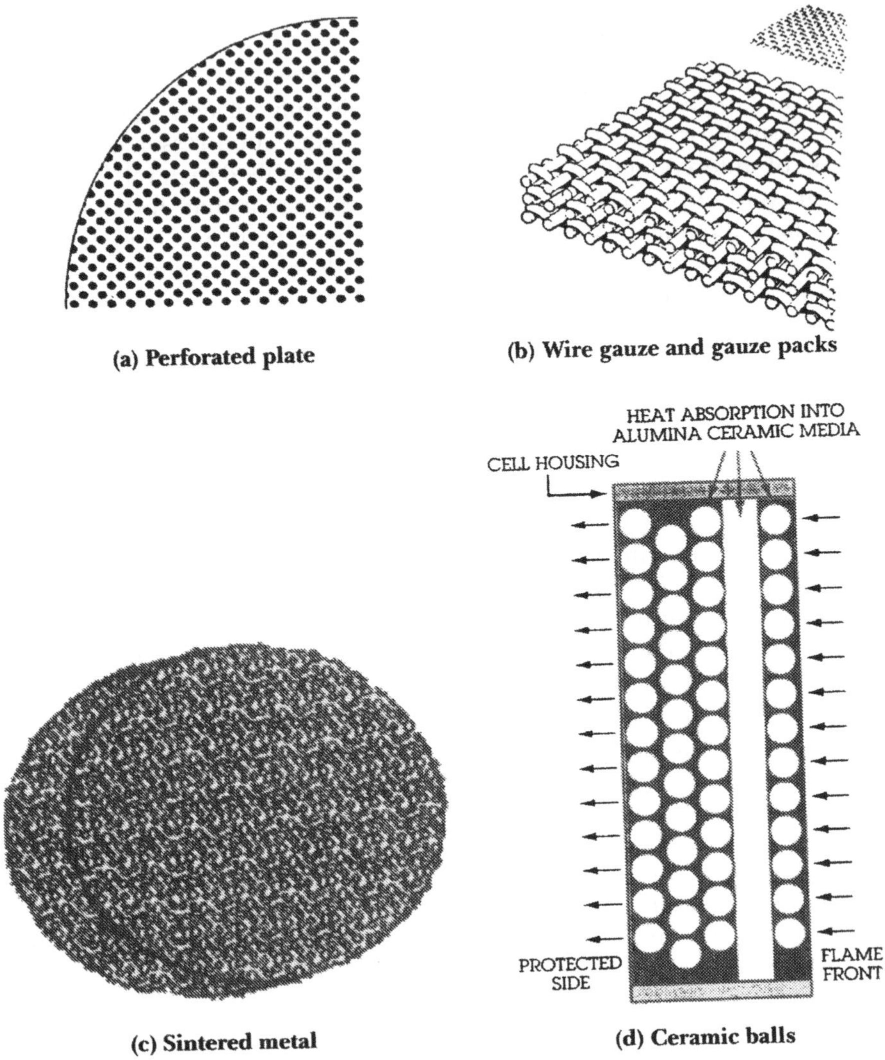

FIGURE 5-4. Various other flame arrester elements (matrixes) (Sources: HSE 1980, Tornado Flare Systems.)

5.2.9. Metal Shot

These arresters consist of a tower or housing filled with various sizes of metal shot (balls) in about nine zones. The size of the balls varies from 4 to 7 mesh for the larger balls and 40 to 60 mesh for the smallest balls. The larger balls are arranged in the outer layer of a zone and the smaller balls are in the inner layers. A typical unit size is 6 inches OD by 15 inches long, with ¾-inch connections. The size of the apertures depends on the diameter of the shot or balls, which are packed tightly together within the container to prevent movement. One advantage of this flame arrester is ease of assembly and disassembly for cleaning purposes. Another advantage is that it can be made sufficiently robust to withstand detonations. Linde (now Praxair) has a design for nickel shot contained in a thick-walled housing, which has been used to successfully stop acetylene detonations at initial pressures from 15 to 400 psig (Dickerman 1999)

However, disadvantages include weight, a relatively high resistance to gas flow, and the size of the apertures is not directly controlled. Movement of the shot or balls during a deflagration or detonation could lead to failure of the flame arrester (HSE 1980).

5.2.10. Hydraulic (Liquid Seal) Flame Arrester

General

While all the flame arrester types discussed above have a solid arresting element (matrix), the hydraulic (liquid seal) flame arrester contains a liquid, usually water, to provide a flame barrier. It operates by breaking up the gas flow into discrete bubbles by means of an internal device to quench the flame. A mechanical nonreturn valve (check valve) is sometimes incorporated to prevent the displacement of liquid during or after a flame event (deflagration or detonation).

This arrester is usually designed to be effective in one direction only. However, hydraulic arresters exist that are reported to be effective in preventing flame propagation in both directions. Tests to establish this on a particular hydraulic arrester design are described by Flessner and Bjorklund (1981).

Proper design against flashback should ensure mechanical integrity of the vessel and internals during the flame event and prevent loss of the liquid seal. Suitable testing should also be performed to ensure that a hydraulic flame arrester design will work for a specific application. Testing procedures are provided in the new CEN standard (CEN 2001) for hydraulic flame arresters. See Section 5.3.18 for recommendations on monitoring and instrumentation for liquid level assurance.

API RP 521 (1997) discusses the design of hydraulic flame arresters (liquid seal drums) for flares. Figure 5-5 shows a typical flare stack seal drum. There are some uncertainties about the effectiveness of the API

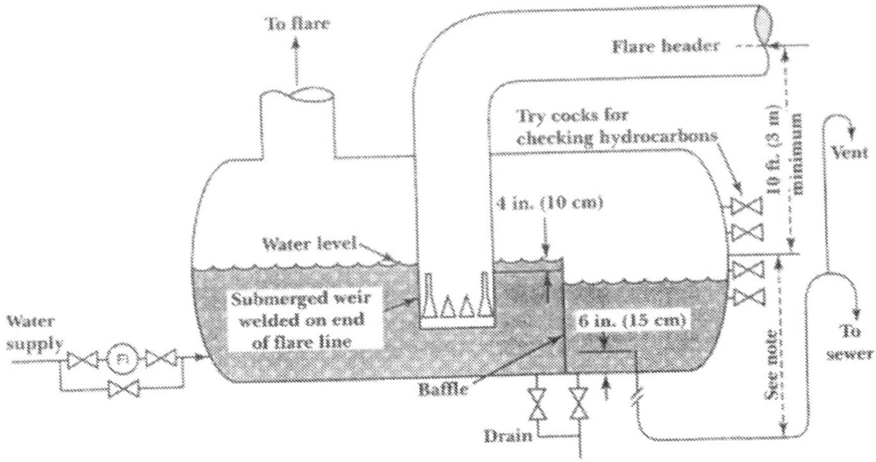

Note: The sewer seal should be designed for a minimum of 175 percent of the drum's maximum operating pressure.

FIGURE 5-5. Sketch of a typical API flare stack seal drum. (Source: API representative 521, Appendix D. Reprinted courtesy of the American Petroleum Institute.)

design for handling detonations and even deflagrations, and a revision of the section on flare seal drums is under way (Straitz 1999).

Hydraulic flame arresters are most commonly applied in large pipe diameters where fixed-element flame arresters are either cost prohibitive or otherwise impractical (e.g., very corrosive gas). This arrester is bulky and requires that the liquid level be maintained either automatically or by regular inspection. A low-liquid-level sensor and alarm are recommended. At low liquid level this arrester might fail, and if the seal liquid is lost, there is no effective barrier to flame propagation. One advantage is that it is not prone to blocking by dirt or other solids collected in the seal liquid. However, it is essential that the liquid used does not react with the gas components and that appropriate measures are taken to prevent freezing. Freeze protection can be provided by using a seal fluid with an antifreeze added, or a liquid that does not ordinarily freeze such as mineral oil. Heat tracing is more commonly employed than antifreeze solutions in many refineries and petrochemical plants. Also, a problem may be caused by foaming agents (Britton 1996). The choice of the seal fluid should consider factors such as compatibility with the process gases, potential scaling, corrosion, or other fouling phenomena.

Hydraulic flame arresters may fail to stop high flame speed gas mixtures under certain conditions. Fundamental test work (Overhoff et al. 1989) demonstrates mechanisms whereby liquid seal arresters may fail to prevent flashback even if gas streams are broken up into discrete, small bubbles. The mechanisms are particularly valid for gas mixtures of high burning velocity, such as hydrogen–air or hydrocarbon–oxygen. Ignition transfer can occur between adjacent bubbles without contact due to hydrodynamic jet effects. The jets occur upon rapid collapse of bubbles of burned gas in the vicinity of discontinuities, which may be adjacent bubbles (the jet effect is analogous to cavitation that produces "jet" erosion at discontinuities at ship propellers). The high velocity hydrodynamic jet may produce compression-ignition of an adjacent bubble, and this process may be transmitted. Alternatively, more closely spaced bubbles might transfer ignition via jets of hot gas, or in the limiting case of a very high void fraction, via direct flame transfer. Several novel designs of liquid seal arresters have been suggested by Overhoff et al. (1989) to mitigate ignition transfer through sparged bubble streams.

Borger et al. (1991) have presented information on a development program on hydraulic flashback protection undertaken at Bayer AG in Germany. The purpose of this program was (1) the development of know-how on hydraulic seals, (2) design of an improved hydraulic seal based on the research performed, and (3) testing of this hydraulic seal on an industrial scale to demonstrate its operation. The paper discusses the results of small-scale tests, which include clarification of the physical phenomena involved in flashback, and some tests on flashback with long time burning in the ethylene–air system.

It is important to realize that due to their size and nature of operation hydraulic flame arresters cannot be readily tested. The vendor should be consulted for examples of successful operation in similar service.

Proprietary Designs

A number of proprietary hydraulic arrester designs are available commercially and are described below.

LINDE HYDRAULIC VALVE ARRESTER

This arrester was developed by the Linde Division of Union Carbide Corporation in the early 1930s. It has been extensively tested during the development of standards and specifications for piping and equipment employed in the handling of acetylene. These tests have repeatedly confirmed the effectiveness and reliability of this arrester. Sutherland and Wegert (1973) have reported on its successful stopping of an acetylene decomposition. Flessner and Bjorklund (1981) have also described tests done with a Linde hydraulic flame arrester using propane as the gas. Five test firings were made on the Linde hydraulic flame arrester, and the deto-

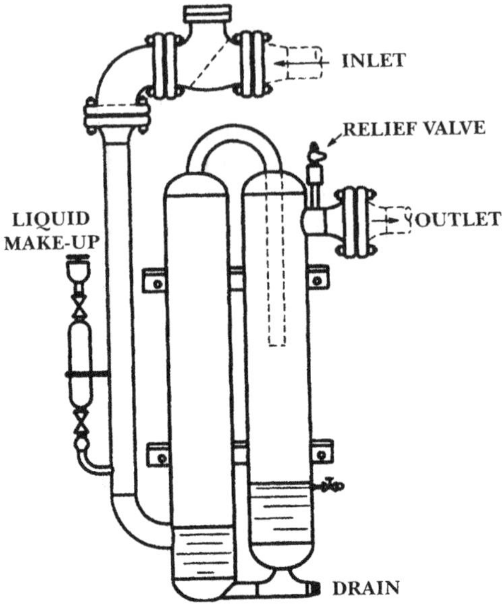

FIGURE 5-6. Linde hydraulic valve flame arrester. (Source: CCPS 1993.)

nation flame was quenched in all cases with no measurable downstream peak pressure pulse.

This flame arrester is no longer made by Praxair (the successor to Linde), but it is available from ESAB Welding & Cutting Products of Florence, SC. It is available in designs for acetylene and fuel gas. For acetylene it can be purchased for handling gas at a maximum inlet pressure of 15 psig and capacities from 500 to 6000 CFH (at 15 psig). For fuel gas, it is available in units at maximum pressures of 20 to 125 psig and capacities from 1000 to 6000 CFH (at 15 psig). Figure 5-6 is a sketch of the Linde Hydraulic Valve arrester.

JOHN ZINK BUBBLE-SCREEN LIQUID SEAL FLAME ARRESTER

In this arrester the gas flows into the seal liquid through a dip-pipe that passes vertically downward through the gas space, and then exits through the seal tip (also called seal head). This seal tip is a distributor that disperses the gas through the liquid as fine bubbles. The latest design has a perforated conically shaped seal tip with the holes facing upward (see Figure 5-7). According to the John Zink Company, this liquid seal flame arrester has been used successfully in a number of applications such as in flare systems, gasoline terminal operations (e.g., tank truck, ship, and barge filling) and even for acetylene and ethylene oxide (EO) gases.

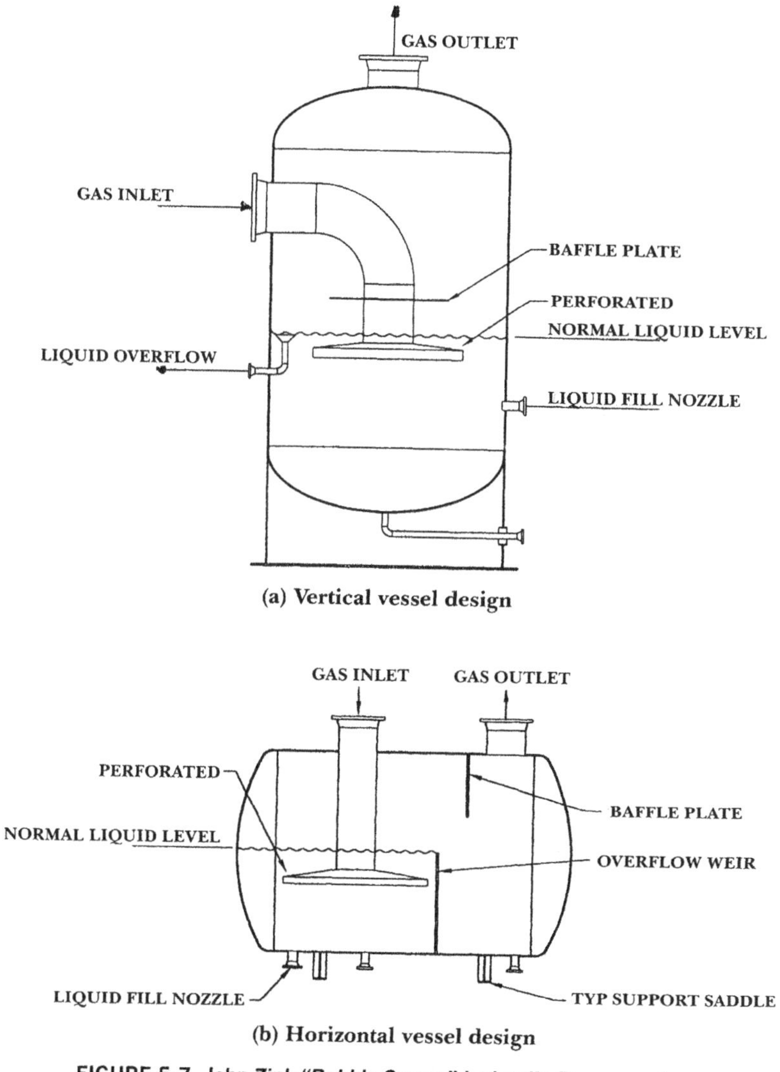

FIGURE 5-7. John Zink "Bubble Screen" hydraulic flame arrester. (Source: John Zink Company.)

NAO INC.

NAO Inc. has a number of proprietary designs for both vertical and horizontal vessel hydraulic flame arresters. Figure 5-8 shows the main components of a vertical dual feed hydraulic flame arrester, and Figure 5-9 shows the details of a horizontal hydraulic flame arrester.

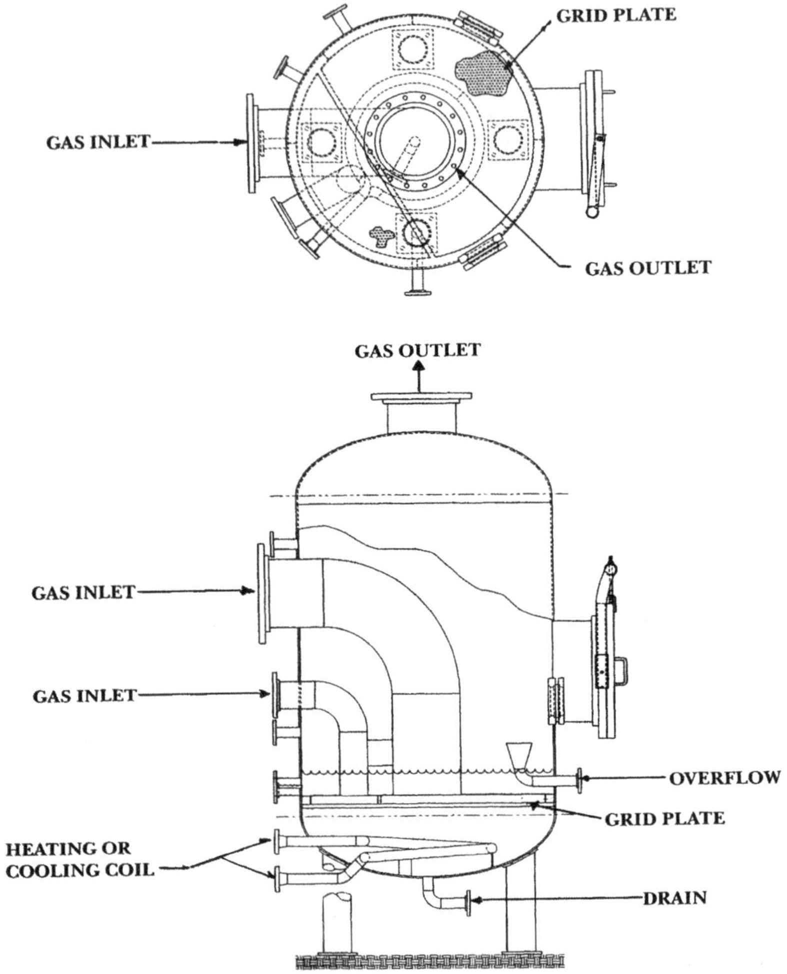

FIGURE 5-8. NAO vertical hydraulic flame arrester. (Source: NAO, Inc.)

The NAO design uses a perforated (bubbler) plate with a skirt and bypass gap in case the bubble holes (about ¼-inch in diameter) get plugged. The design includes a minimum of 6 inches of liquid seal above the bubbler plate, and the gas superficial velocity is limited to 1 to 3 ft/s.

NAO has successfully tested hydraulic flame arrester designs for detonations of hydrogen and oxygen (Mendoza 1999). The NAO hydraulic arresters also have an internal detonation inhibitor (shock absorber) upstream of the gas exit nozzle. See the article by Overhoff et al. (1989) for discussion of shock effects in hydraulic flame arresters.

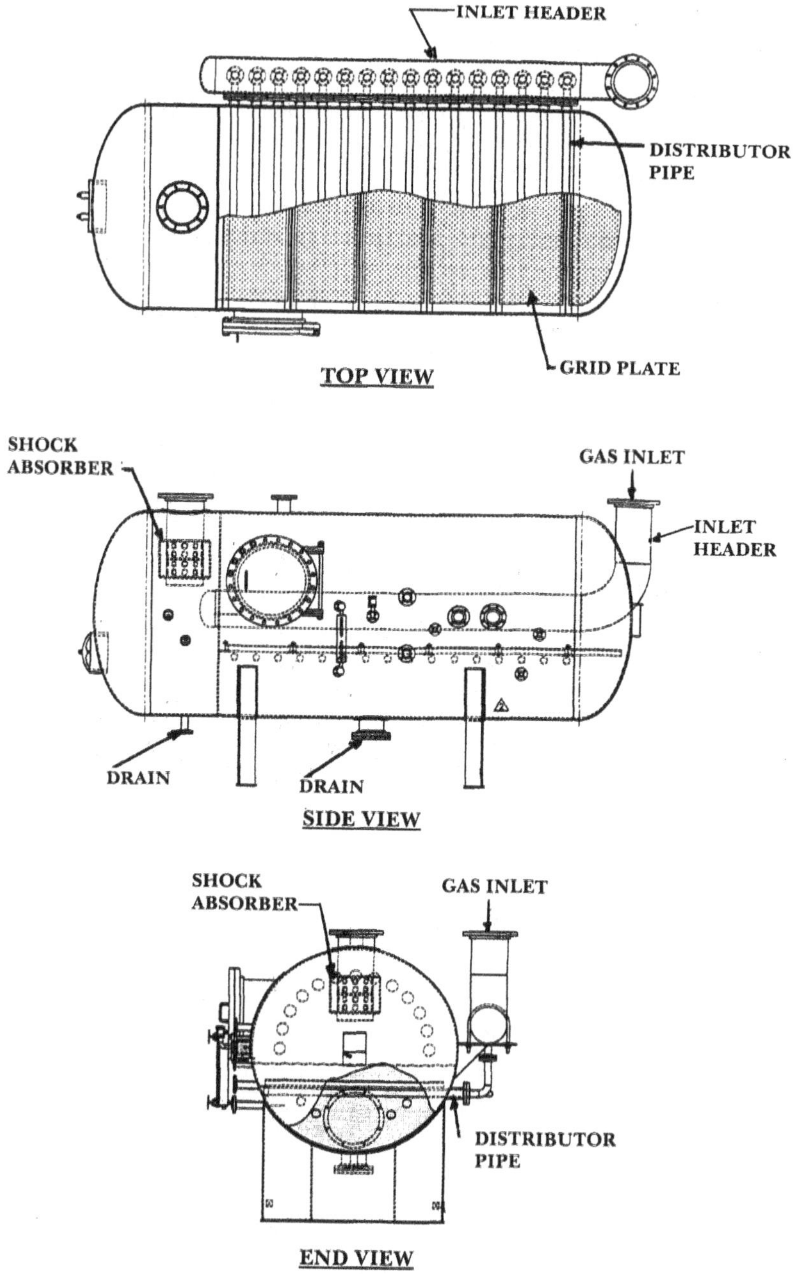

FIGURE 5-9. NAO Horizontal hydraulic flame arrester. (Source: NAO, Inc.)

PROTEGO™

Protego™ (Braunschweiger Flammenfilter GmbH) also has proprietary designs for both vertical and horizontal hydraulic flame arresters. These flame arresters are designed with an internal shock absorber for protecting against detonations and for long-burning situations. The gases are introduced beneath the seal liquid by means of a series of perforated sparger pipes. These units are routinely provided with level and temperature instrumentation, and automatic seal liquid makeup controls. A quick-clos-

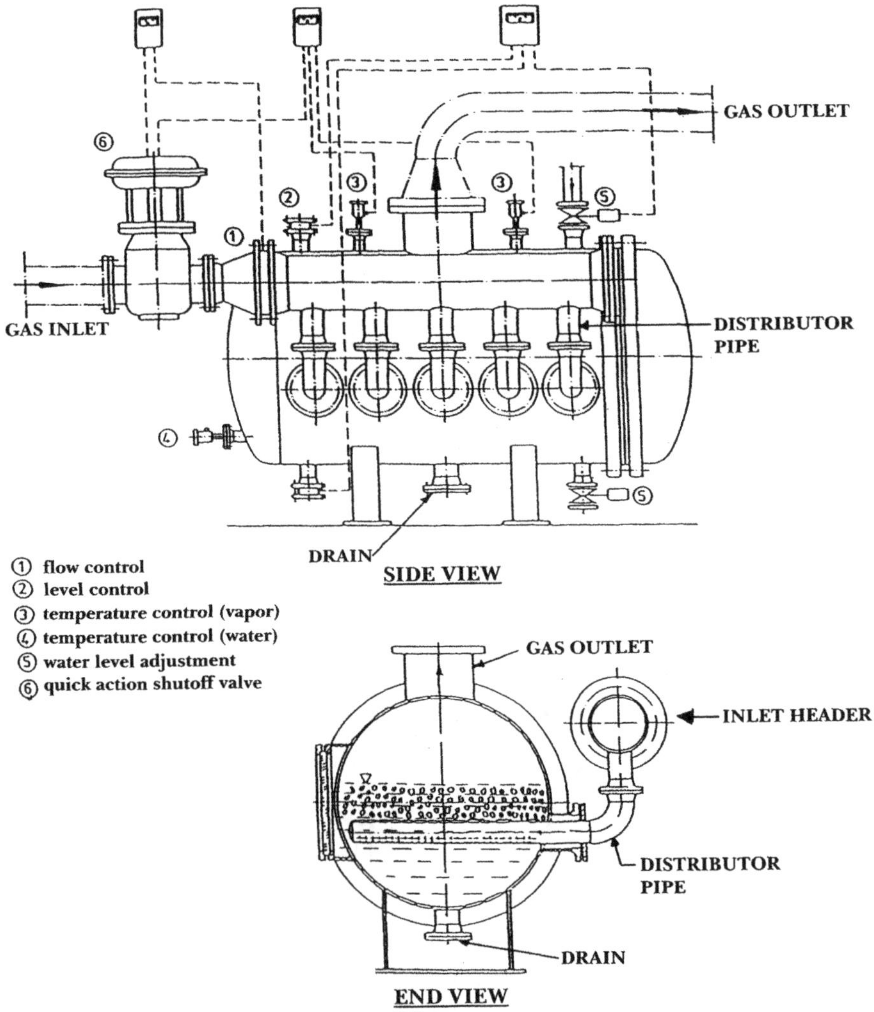

FIGURE 5-10. Protego™ horizontal hydraulic flame arrester. (Source: Portego/Braunschweiger Flammenfilter GmbH)

ing valve at the inlet of the gas-entry manifold may also be provided. Figure 5-10 is a schematic of a Protego™ horizontal hydraulic flame arrester and associated control instrumentation. Test data for this hydraulic flame arrester design are available from Protego™.

5.2.11. Packed Bed Flame Arrester

Flame arresters consisting of a tower, or other container, filled with pebbles, Raschig rings or other packings, have been used for many years with success. The sizes of the apertures available for flame quenching depends on the sizes of the pebbles or packings, and the effectiveness of the arrangement is usually increased by wetting the packing with water or oil. The advantages of this arrester are that it is easily dismantled for cleaning and reassembled, and that it can be made sufficiently robust to withstand severe explosions. It has the following disadvantages: it may be large, it has a relatively high resistance to gas flow, and the size of the passages through the arrester is not directly controlled. Also, movement of the packing during an explosion could lead to failure of the arrester to quench a flame.

The packed bed tower arrester has been used successfully for many years for systems handling acetylene at low and medium pressures. The design of packed bed arresters is discussed in CGA Pamphlet G-1.3 (1970). Standard American practice is to use 1-inch metal (carbon steel or stainless steel) Raschig rings of minimum 20 gauge wall thickness. In general, the packing ring size needs to be decreased as the acetylene pressure is increased, the largest being typically 25mm and the smallest 10 mm (Britton 2000). The recommended packed height of a liquid-wetted arrester is a minimum of 4 feet. For a dry packed tower, it is recommended that the packed height be doubled. For sizing the tower diameter, CGA Pamphlet G-1.3 recommends a superficial velocity of 2 ft/s or less, and a tower diameter not less than 15 times the diameter of the packing. Further information on packed bed arresters for acetylene service is presented by Saacke (1963).

Flessner and Bjorklund (1981) have reported on flame arrester tests on packed beds of 1-inch aluminum Pall rings. Five tests were made using gasoline–air mixtures, and the detonation flame was arrested in all tests, and there were no measurable downstream pressure pulses. Flessner and Bjorklund also discuss tests by other investigators. Bjorklund and Kushida (1982) have also reported on tests with 1-inch aluminum Pall rings with single 30-mesh stainless steel screen retainers. This packed tower arrangement was effective in arresting flashback flames (deflagrations) from tests with propane, ethylene, and gasoline vapor–air mixtures. However, the packed bed of aluminum Pall rings without the single 30-mesh screen retainer was not effective in arresting flashback flames from gasoline vapor–air mixtures in three out of three tests; therefore the other fuel–air mixtures were not tested.

5.2.12. Velocity Flame Stopper

A velocity flame stopper is a special type of flame arrester used only in end-of-line applications. It usually consists of a tee with holes in it (see Figure 5-11). Velocity flame stoppers function only when the flames arrive at the flame stopper face from the downstream side with respect to the direction of gas flow through the holes. This arrester only stops deflagrations, not detonations. Therefore, it cannot be used as an in-line arrester. It operates on a principle quite different from other arresters. That is, the velocity of the upstream gas passing through the arrester must be sufficiently high enough to prevent a flame from propagating through the arrester from the downstream side. The velocity flame stopper principles and design are discussed by Howard (1982).

The hole size used is larger than that necessary to quench a flame in a stagnant flammable gas mixture, i.e., larger than the quenching diameter. Howard recommends that the velocity necessary to prevent flashback be calculated by the following equation:

$$u_T = 0.2015 g_L D \tag{5-1}$$

where u_T is the turbulent flashback velocity (m/s), g_L is the laminar velocity gradient at the pipe wall below which flashback can occur (sec^{-1}), and D is the inside diameter of the pipe (m). The parameter g_L is also called the critical boundary velocity gradient (Grumer et al. 1956), and is a function of a specific gas and its concentration. It tends to have a maximum value at a concentration somewhat above the stoichiometric value. Howard (1982)

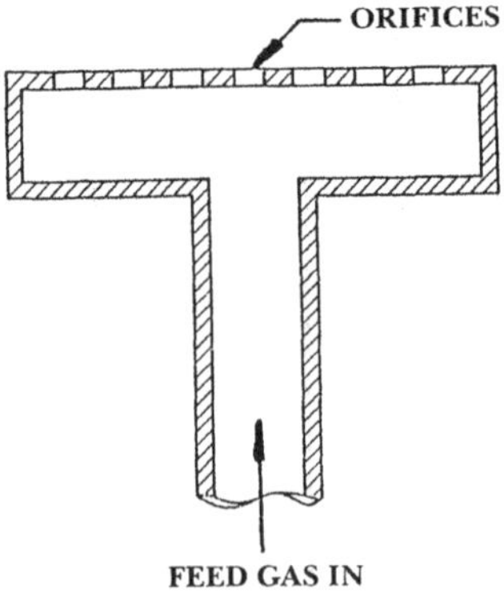

FIGURE 5-11. Sketch of a velocity flame stopper. (Source: Howard 1982.)

lists some values for g_L for methane, ethane, propane, ethylene, propylene, and hydrogen. Grumer et al. (1956) present g_L data for some two-component fuels (methane–hydrogen, carbon monoxide–hydrogen, methane–carbon monoxide, propane–hydrogen, ethylene–hydrogen, nitrogen–hydrogen, and some other mixtures). Howard recommends that for normal design the minimum velocity through the holes should be four times the turbulent flashback velocity calculated by Eq. (5-1). From the standpoint of stopping flames, there is no limit to how small the holes may be made. It is not known how large the holes can be made for a fully functional velocity flame stopper, but holes as large as 2 inches have been used in commercial installations (Howard 1982).

A velocity flame stopper is effective only as long as there is a sufficient gas flow through it. If gas stream can be subject to low flow deviations during normal or upset operating conditions, a highly dependable auxiliary gas flow must be provided. The reliability of this auxiliary gas system will affect the selection of the velocity flame stopper.

Velocity flame stoppers have been used for feeding waste fuel gas to furnace burners when the gas can become flammable due to contamination with air. They have also been used for feeding waste or depleted air streams to furnaces when the air streams can become contaminated with flammable gases (Howard 1982). It should be noted that a furnace pressure transient may render this device ineffective and consideration should be given to providing an upstream detonation flame arrester. In this arrangement a demand will only be placed on the detonation flame arrester when the velocity flame stopper fails. Therefore, detonation flame arrester maintenance should be minimal.

See Section 5.4.1 for additional information on the use of velocity flame stoppers for hydrogen service.

5.2.13. High Velocity Vent Valve

A high velocity vent valve is used primarily at the outlet of a vent pipe on a flammable liquid cargo tank of a seagoing tanker or barge. The vent valve contains either a weighted flap or a weighted disk that adjusts the opening available for flow in accordance with the pressure at the inlet of the valve in such a way that the efflux velocity cannot be less than 30 m/s. The jet of flammable vapors is ejected into the atmosphere, and if the jet ignites, the jet velocity is so high that the flame cannot flash back into the tank.

Schampel and Steen (1975) describe experimental equipment and tests carried out at the Physikalisch-Technische Bundesanstalt (PTB) in Germany on high velocity vent valves. Also, conditions for a sufficient air entrainment and dilution of the vented flammable vapors are discussed.

5.2.14. Conservation Vent Valves as Flame Arresters

NFPA 30 recognizes that a conservation vent valve (pressure-vacuum valve) is an alternative to a flame arrester under certain circumstances. This recognition is based on tests begun in 1920 and is supplemented by many years of experience.

Even with highly flammable vapors, flame cannot pass back through an opening if the efflux velocity exceeds some critical value. Tests have demonstrated that the critical velocity is approximately 10 ft/s for mixtures of gasoline components and air issuing from openings typical of tank vents. Flame arresters are not considered necessary below a conservation vent valve on a storage tank provided the valve is set to close when the tank pressure falls below ¾-inch water gauge, and the discharge is not through a piping system in which a detonation can occur (API RP 2210 2000). Under these conditions the gas velocity through the valve will be considerably greater than the speed at which the flame can propagate past it into the tank. To address the possibility of airborne "sparks" (such as hot cinders) being drawn through the vent without being quenched, the USCG requires that a tested flame screen be installed on the vacuum port (33 CFR Part 154, Subpart E, Section 154.814).

Tests have also shown that under some circumstances a long-burning flame at the valve outlet could damage the valve sufficiently to interfere with its closing. Under such circumstances, flashback may occur when the flow rate falls below the critical velocity if a flammable mixture exists inside the tank (API RP 2210 2000). It is pointed out that a long vent line from a conservation vent may result in flame acceleration and possibly detonation resulting in flame passage into the tank (Britton 1996). Where a long vent line is necessary, a detonation flame arrester should be installed.

5.3. Selection and Design Criteria/Considerations

The concepts of the National Electrical Code (NEC) groups and the Maximum Experimental Safe Gap (MESG) are important criteria in the selection and specification of dry type flame arresters. These are explained below.

5.3.1. Classification According to NEC Groups and MESGs

For flame arrester selection, gases are classified according to two methods: National Electrical Code (NEC) groups or the Maximum Experimental Safe Gap (MESG).

The NEC group method for classifying flame arresters is similar to that used for electrical area classification. NFPA 497 (1997) provides the criteria for classifying gases into NEC groups for suitability for electrical area classification.

Originally the classification of materials was derived from tests of proprietary explosion-proof (flameproof) enclosures. There were no published criteria. Equipment was approved relative to the lowest ignition temperature of any material in the group (Magison 1987). In about 1965 the U.S. Coast Guard asked the National Academy of Sciences (NAS) to form a panel to classify 200 materials of commerce. The Electrical Hazards Panel of the Committee on Hazardous Materials was formed by the NAS. The Panel studied many ways to estimate the hazard classification of materials. The Panel finally reported to the U.S. Coast Guard in 1970 that no workable, predictive scheme could be defined, and it then proceeded to assign tentative classifications to the 200 materials.

Classification considered a number of factors like similarity of chemical structure, flammability characteristics such as the MESG, the minimum igniting current (MIC) ratio or the minimum ignition energy (MIE), and the hazard level assigned by other authorities. The Panel recommended testing of 21 compounds in the Westerberg explosion test vessel at Underwriters Laboratories, Inc. to provide reference MESG data. Tests were also performed on an additional 11 compounds. Finally, in 1975, the Panel issued its final report to the Coast Guard. In 1982 the National Materials Advisory Board (NMAB) issued a report containing classification data on 1500 gases and vapors and 350 dusts (NMAB 1980). This report has been used for many years for classification of explosion-proof electrical equipment, even though some of these classification data are not based on experimental values but are based on engineering judgment.

More recently, the NFPA has used the MESG and the MIC ratio for classifying explosion-proof electrical equipment (NFPA 497 1997), and this approach can also be used for classifying flame arresters. In this method, NEC Class I combustible materials (vapors or gases) are divided into four groups:

Group A: Acetylene

Group B: Flammable gas, flammable liquid-produced vapor, or combustible liquid-produced vapor mixed with air that may burn or explode, having either a maximum experimental safe gap (MESG) value less than or equal to 0.45 mm or a minimum igniting current ratio (MIC ratio) less than or equal to 0.40. Typical Class I, Group B gases are gases containing more than 30% hydrogen by volume, butadiene, ethylene oxide, propylene oxide, and acrolein.

Group C: Flammable gas, flammable liquid-produced vapor, or combustible liquid-produced vapor mixed with air that may burn or explode, having either a MESG value greater than 0.45 mm and less than or equal to 0.75 mm, or a MIC ratio greater than 0.40 and less than or equal to 0.80. Typical Class I, Group C gases are ethylene, ethyl ether, and other gases of equivalent hazard.

Group D: Flammable gas, flammable liquid-produced vapor, or combustible liquid-produced vapor mixed with air that may burn or explode, having either a MESG greater than 0.75 mm or a MIC ratio greater than 0.80. Typical Class I, Group D gases are methane and other alkanes, alcohols, acetone, benzene, and other gases of equivalent hazard.

In Europe, rather than Groups A through D, gases and vapors are classified in Groups IIA through IIC. A comparison of the U.S. and European groupings is as follows:

U.S.	Europe
Group A	IIC
Group B (Part)	IIC
Group C	IIB
Group D	IIA

The International Electrotechnical Commission (IEC) has placed hydrogen, acetylene, carbon disulfide, and ethyl nitrate into Group IIC. The United States, on the other hand, has separated hydrogen and acetylene into different groups and does not classify carbon disulfide.

MESG is defined in terms of the precise test method and apparatus used, of which there are three variants: British, IEC, and Underwriters Laboratories, Inc. Each apparatus consists of two combustion chambers connected by a slot of specified size and variable width. The separate chambers are filled with the test mixture. The MESG is the maximum slot width that prevents flame propagation between the chambers for all compositions of the test gas in air under the specified test conditions. Phillips (1987) describes and compares these three types of experimental apparatus for determining the MESG.

The MESG is used in the USCG standard to compare gases for detonation flame arrester applications, under the assumption that flames of mixtures with smaller MESGs are harder to stop. This assumption has not yet been verified by comprehensive flame arrester tests, although related work by Frobese and Forster (1992) found that "the MESG is indeed a suitable ordering and evaluating parameter, independent of the specific fuel" for evaluating detonation processes at branches in piping networks.

The MESG cannot be determined theoretically and has to be measured experimentally. The experimental measurement of the MESG suffers from a strong apparatus dependency. The 20-ml vessel adopted by the IEC can yield results that are quite different from those obtained from the UL-Westerberg apparatus. Also, for the same apparatus, different test conditions give different results for the MESG. For example, changing the location of the ignition source in the test vessel, which affects the explosion pressure developed in the test chamber, may lead to different values for the MESG. Strehlow et al. (1979) and Phillips (1981) have attempted to explain the reasons for these differences in the test data.

MESG values for gas mixtures can be tentatively estimated using the relationship of Le Chatelier as recommended by Britton (1996) and illustrated in Appendix B of NFPA 497 (1997). A modified Le Chatelier rule was first proposed by the NFPA 497 Committee for estimation of the MESG of fuel mixtures (excluding acetylene) for electrical area classification. This included the effect of inerts, unlike the original Le Chatelier rule. The Committee used unpublished UL data to justify using this rule (Briesch 2000). Britton (1996) originally proposed using this approach for MESG estimation when selecting detonation flame arresters. More recently, however, Britton (2000a) has reassessed the use of the MESG for selecting flame arresters, and has recommended that this approach for estimating multicomponent MESG values not be used until further verification.

The use and possible misuse of the MESG criterion for selecting flame arresters were discussed by Britton (2000a). His conclusions were as follows:

1. Care is needed in applying the concept of MESG to selection of DDAs. There have been no systematic studies proving that DDA performance can be directly correlated with MESG. For instance, there should be an interaction with AIT, which is independent of MESG and is also relatively insensitive to the concentration of inert components. If a gas mixture has a low AIT, reignition might occur in the large, compressed volume on the protected side of a DDA, especially under restricted end deflagration (RED) conditions. It should be noted that current RED test protocols do not require optimization with respect to either gas composition or ignition location. Test results have only been reported for gases with relatively high AITs, such as ethylene and propane.
2. If a gas stream contains inert components and the MESG is estimated using the combination formula in NFPA 497, the result will be selection of an arrester element having larger openings than would be required for the flammable components alone. For example, a propane-type rather than an ethylene-type arrester might be

selected. This approach assumes not only that DDA performance correlates directly with MESG, but that it is appropriate to take credit for inert gas stream components. In practice, the concentration of inerts is often decreased by the very upset conditions that causes the DDA to be challenged in the first place, for example, condensation of steam in a pipe or substitution of air for inert padding gas in a tank.

3. Assuming that the MESG approach to DDA selection is an oversimplification, a small-scale test specifically addressing gas properties with respect to DDA performance might need to be developed. Meanwhile, it is suggested that Westerberg MESG data be used to rank gas sensitivities for DDA selection even though IEC data are more appropriate for electrical equipment design. However, few Westerberg data have been published and the test is relatively costly. In practice, the choice of DDAs is typically limited to only three types, reflecting certification using propane, ethylene, or (in rare cases) a more sensitive gas such as hydrogen. The designer might only have MESG data measured in the IEC apparatus, or an estimated value in the case of mixtures. If there is a choice between propane- and ethylene-type DDAs and the MESG (≥0.9 mm) indicates the propane-type, yet there is reason to believe that a Westerberg test would produce a MESG of about 0.8 mm or less, the designer might consider using the ethylene-type DDA. More quantitative guidance based on full-scale DDA testing is desirable.
4. It is proposed that a collective effect be made to investigate the relevance of MESG in selecting DDAs. A candidate measure of performance is the DDA acceptance pressure determined under optimized RED conditions. This can be determined as a function of calculated MESG for mixtures that include high and low AIT gases (such as ethane plus n-hexane), high and low MESG gases (such as ethane plus hydrogen), plus an inert gas such as nitrogen.

Table 5-1 lists MESG values published from four different sources (Britton 2000a). The USCG values are taken from Attachment 1 to Appendix B of Part 154 of 33 CFR. The Westerberg values are from a report by the US National Academy of Sciences (NAS 1975). The British values are also from the NAS report. The NFPA values are from NFPA 497. Although MESG values are listed to the nearest 0.01 mm, this does not reflect measurement accuracy. The minimum gap width measurements used in MESG testing are 0.025 mm (Westerberg apparatus) and 0.02 mm [European (IEC and British) apparatus]. Repeatability data are unavailable for either the Westerberg or British apparatus. Repeatability and "round robin"

TABLE 5.1
Comparison of Published MSEG Values (Britton 2000a)

Chemical	Coast Guard MESG (mm)	Westerberg MESG (mm)	British MESG (mm)	NFPA 497 MESG (mm)	AIT (°C)
methanol	0.92	n/a	0.91	0.92	464
ethanol	1.02	n/a	1.02	0.89	363
n-butyl alcohol	0.94	n/a	0.94	0.91	343
iso-butyl alcohol	0.96	n/a	0.96	n/a	405
n-amyl alcohol	0.99	n/a	0.99	n/a	300
allyl alcohol	not listed	0.66	n/a	n/a	378
acetaldehyde	**not listed**	**0.43**	**n/a**	**0.92**	**175**
n-butyraldehyde	**not listed**	**0.43**	**n/a**	**0.92**	**218**
acrolein	not listed	0.46	n/a	n/a	220
crotonaldehyde	**not listed**	**0.48**	**n/a**	**0.81**	**232**
methane	1.17	1.12	1.17	1.12	537
ethane	not listed	n/a	n/a	0.91	472
propane (commercial)	0.96	0.94	0.96	0.97	450
butane (commercial)	1.07	0.99	1.07	1.07	varies
n-pentane	**1.02**	**0.56**	**0.99**	**0.93**	**260**
n-hexane	0.96	n/a	0.96	0.93	225
n-heptane	0.96	n/a	0.91	0.91	204
octanes	1.04	n/a	1.04	0.94	varies
cyclohexane	0.94	n/a	0.94	0.94	245
diethylamine	**not listed**	**0.69**	**n/a**	**1.15**	**312**
triethylamine	not listed	0.53	n/a	n/a	249
ethylenediamine	not listed	0.74	n/a	n/a	385
ethyleneimine	not listed	0.61	n/a	n/a	320
pyridine	not listed	0.84	n/a	n/a	482
benzene	0.99	n/a	0.99	0.99	498
p-xylene	1.07	1.22	n/a	n/a	528
mixed xylenes	1.07	n/a	1.07	n/a	varies
styrene	not listed	0.94	n/a	n/a	490
methyl acetate	0.99	n/a	0.99	0.99	454
ethyl acetate	1.04	n/a	1.04	0.99	426
n-propyl acetate	1.04	n/a	1.04	1.05	450
n-butyl acetate	1.02	n/a	1.02	1.04	425
iso-amyl acetate	0.99	n/a	0.99	n/a	n/a
vinyl acetate	not listed	1.04	n/a	0.94	402
ethyl acrylate	not listed	0.81	n/a	0.86(a)	372
diethyl ether	**0.86**	**0.30**	**0.84**	**0.83**	**160**
di-isopropyl ether	not listed	n/a	n/a	0.94	443
ethylene oxide	**0.65**	**0.33**	**0.66**	**0.59**	**445**
propylene oxide	**not listed**	**0.53**	**n/a**	**0.70**	**449**
tetrahydrofuran	**not listed**	**0.61**	**n/a**	**0.87**	**321**
morpholine	**not listed**	**0.43**	**n/a**	**0.95**	**290**
epichlorohydrin	not listed	0.56	n/a	n/a	411
ethylene dichloride	not listed	0.99	n/a	n/a	413
vinyl chloride	**not listed**	**0.74**	**0.96**	**0.96**	**472**
acetone	1.02	n/a	1.02	1.02	465
methyl ethyl ketone	1.02	n/a	0.84	0.84	404
mesityl oxide	not listed	0.84	n/a	n/a	344
acrylonitrile	not listed	0.81	n/a	0.87	481
ethylene	0.71	0.69	0.69	0.65	450
propylene	not listed	0.89	n/a	0.91	455
1,3-butadiene	not listed	0.79	0.84	0.79	420
isoprene	not listed	0.86	n/a	n/a	395
di-isobutylene	not listed	0.81	n/a	n/a	391
acetylene	**≤0.02**	**0.08**	**0.25**	**0.25**	**305**
ammonia	**3.33**	**1.02**	**3.18**	**3.17**	**651**
carbon disulfide	**0.20**	**0.05**	**0.20**	**0.20**	**90**
carbon monoxide*	**0.92**	**0.63**	**n/a**	**0.94(a)**	**609**
hydrogen	**0.10**	**0.08**	**0.28**	**0.28**	**520**
hydrogen sulfide	**not listed**	**0.66**	**n/a**	**0.90**	**260**

(a) = Value not currently listed in NFPA 497; corrected MSEG taken from Lunn (1982a, b).
n/a = Value not available
* = Humidified to provide a source of hydrogen (allowing faster combustion).
Chemicals exhibiting a difference of at least 0.1 mm between Westerberg and European values appear in bold.
Additional MSEG values can be found in Lunn (1982 a,b).

reproducibility data for the IEC apparatus are also unavailable. European (British) data reported by Lunn (1982a) were corrected to standard conditions of 1 atm and 20°C using empirical formulas. For less volatile chemicals, this could increase the observed MESG values by up to about 0.1 mm. Westerberg data were not corrected using this method. Differences of at least 0.05 mm between Westerberg and European data are therefore probably insignificant. Chemicals exhibiting a difference of at least 0.1 mm between Westerberg and European MESG values are highlighted in bold. Autoignition temperatures are taken primarily from NFPA 325 (1994).

As can be seen from this table, the MESG values for a specific substance are quite often different depending on the source, due to the use of different experimental apparatus. The most notable difference is in the case of acetylene, whose USCG value is more than an order of magnitude smaller that that listed for the British or NFPA 497 value. The MESG values cited in the USCG Regulations for Marine Vapor Control Systems (33 CFR Part 154, Subpart E) are primarily taken from IEC Standard 79-1A (1982). The table also shows that the AIT is not the only factor governing MESG differences between the Westerberg and European test apparatus. Chemicals such as hydrogen, carbon monoxide, vinyl chloride, and epoxides give smaller MESG values in the Westerberg apparatus despite their high AITs. All of these have unusually wide flammable ranges, implying a fast rate of combustion over a wide range of fuel concentrations. The flammable range is another "gas sensitivity" parameter that might be considered when attempting to identify gases whose Westerberg MESG values are significantly lower than the European MESG values. Since hot gas exiting a DDA approximates to a back-mixed jet with minimal entrainment at its base, neither type of test properly simulates DDA operation. However, owing to the greater confinement produced in its receptor chamber, the Westerberg apparatus should be better able to resolve "gas sensitivity" differences with respect to DDA performance.

5.3.2. Reactions and Combustion Dynamics of Fast-Burning Gases

When using dry type flame arresters for fast-burning gases, such as hydrogen, acetylene, ethylene oxide and other gases with high fundamental burning velocities, small apertures are needed to quench the flame. Mixtures of fast-burning gases and Group D gases may also require small apertures to quench the flame. The MESG of gas mixtures can be estimated using Le Chatelier's rule (see Chapter 9). Dry type flame arresters must be judiciously located to prevent deflagration-to-detonation transition in piping handling fast-burning gases. Hydraulic (liquid seal) flame arresters have been successfully used for quenching flames of fast-burning gases, e.g., acetylene (Sutherland and Wegert 1973) and hydrogen (Rao et al. 1980).

A number of gases may decompose (self-react) and propagate flames in the absence of any oxidant provided that they are above minimum conditions of pressure, temperature, and pipe diameter. Common examples are acetylene, ethylene oxide, and ethylene. Some, like acetylene, can decompose in a detonative manner, while ethylene cannot detonate in the absence of an oxidant, whatever the run-up length (CCPS 1993). Thus, detonation arresters must be used for acetylene, but deflagration arresters may be used for ethylene, even for in-line applications.

5.3.3. Flame Propagation Direction

The flame propagation direction affects the type of flame arrester selected. An end-of-line or in-line deflagration flame arrester used for the protection of an individual tank may be of a unidirectional design because the flame will only propagate from the atmosphere towards the tank interior. A bidirectional flame arrester design, however, is needed for an in-line application in a vapor recovery (vent manifold) system because the vapors must be able to flow from the tank interior into the manifold, or from the manifold into the tank interior. Consequently, flame may propagate in either direction.

5.3.4. Quenching Diameter, Quenching Length, and Flame Velocity

The quenching diameter and quenching length are two very important parameters needed for the design of dry deflagration and detonation flame arresters. This section presents a brief overview of flame quenching and equations that have been proposed for estimating the quenching diameter and quenching length.

Whether a flame is transmitted through a flame arrester depends on the length and aperture size of the arrester, the approach velocity of the flame, the pressure rise, and the temperature of the arrester (Wilson and Flessner 1978). Wilson and Flessner state that the evidence indicates that low-speed flames can be quenched by an array of small passageways placed in a duct, provided that the effective passageway diameter (critical diameter) meets the following criterion:

$$d_{cr} < 30\alpha/S_u \tag{5-2}$$

where d_{cr} is the critical (quenching) diameter (m), α is the thermal diffusivity in air (m²/s), and S_u is the fundamental burning velocity (m/s).

For high-speed flames, which are usually accompanied by a pressure rise, an array of apertures must have sufficient pressure drop to decelerate the flame, and sufficient length to achieve the heat loss needed to quench the flame. Thus, the diameter criterion is not sufficient and the effective length of the passageway must meet the following criterion (Wilson and Attalah 1975):

$$L > 2(S_t)(d_{cr})^2 \qquad (5\text{-}3)$$

where L is the effective passageway length (cm), S_t is the turbulent flame speed in the duct (m/s), and d_{cr} is the critical (quenching) diameter of the passageway (cm). For methane, methyl alcohol, butane, gasoline vapor, and other common gases, these two criteria imply that flame arresters of greater than 1-inch effective length and less than 0.030 inches effective diameter will prevent passage of both low and high-speed flames (Wilson and Flessner 1978).

The article by Wilson and Flessner gives the dividing line as roughly 50 ft/s between "slow" flames that can be simply quenched and "fast" flames that must also be decelerated. "Fast" flames described in the article have speeds above 60 ft/s as opposed to "turbulent" flames, which are described as having speeds from 5 to 100 m/s in most venting systems. The test rig described in the article was composed of 6-inch diameter pipe.

Two factors can make the approaching flame more difficult to quench than a laminar flame in a tube. First, turbulent transport of heat and mass ahead of the flame causes the flame to accelerate to speeds greatly in excess of the range of laminar flame speeds. For most piping systems, the turbulent flame speed ranges from 5 to 100 m/s, which for methane is 10 to 200 times greater than the laminar flame speed (Wilson and Flessner 1978). The turbulent flame speed is larger for high flow velocities and large pipe diameters. Flames of high speed are more difficult to quench than laminar flames because the flame arrester has less time to extract heat from the flame. Second, constrictions in the piping prevent free expansion of the products of combustion, and this leads to a pressure rise. The importance of the pressure rise accompanying a turbulent flame is twofold: (1) the arrester must now satisfy a structural as well as a thermal criterion, and (2) the hot products from the flame may expand through the apertures of the arrester, possibly contributing to ignition.

Quenching Criteria

For a flame to be quenched the flame arrester passageways must be small enough to extract heat from the flame faster than it can be generated by chemical reactions. The surface to volume ratio of flame arresters is important because the flame quenching mechanism depends on heat loss. By forcing the flame to pass through an array of small passageways, the frontal area of the flame is divided into small flamelets, and the array can efficiently extract heat from the sides of a laminar flame. The smaller the cross-sectional area, the greater is the surface to volume ratio, *A/V*, of the flame disk. The rate of lateral heat loss (from the flame sides), Q_{loss}, compared to the rate of heat production by combustion, Q_{chem}, increases in proportion to *A/V*. One approximate measure of the surface to volume ratio is the hydraulic diameter, d_H, defined by:

$$d_H = 4\left(\frac{\text{cross sectional area of passageway}}{\text{perimeter of passageway}}\right) \tag{5-4}$$

It should be pointed out that the hydraulic diameter method does not work well for laminar flow because the shape affects the flow resistance in a way that cannot be expressed as a function of the ratio of cross-sectional area to wetted perimeter (Green and Maloney 1997). However, some flame arrester manufacturers use this method for noncircular flame arrester passages.

Table 5-2 shows the equations for calculating the hydraulic diameter for various flame arrester passageways.

To prevent the passage of a flame an array of apertures must provide sufficient surface area to volume ratio and sufficient length to achieve the necessary quenching. A measure of these two requirements is the ratio of length to diameter of the passageways, *L/D*. For noncircular passageways, the hydraulic diameter, d_H, is the appropriate diameter to use in calculating *L/D* for flame arresters. To calculate the length, the overall length of the arrester (say ½ to l-inch) is multiplied by a tortuosity factor (equal to 1 for tube bundles, parallel plates, and crimped ribbon), which accounts for the bends in the flow path.

It was recognized in the early studies of flame arrester elements that devices that work for laminar flames would require either smaller apertures or longer length for turbulent flames. Wilson and Attalah (1975) have summarized the findings of 14 studies on crimped ribbon, metal foam, tube bundles, perforated blocks, gravel beds, and parallel plates. The best flame quenching performance for the least pressure drop was offered by arresters of the crimped ribbon type; gravel beds and metal foam were also very effective, but they had relatively large pressure drops and were subject to blockage. Many of the studies maintained length at a constant value large enough so that *L/D* was in the range 20–80 for the values of d_H explored. Therefore, a length criterion was met. In these studies, arresters of ½-inch to l-inch length were normally found to be effective for hydrocarbon/air flames (except acetylene) of up to 200 m/s flame speed, provided that the passageway diameter was small enough.

Palmer (1960), Palmer and Tonkin (1963), and Langford et al. (1961) have studied variations in effective length as well as diameter. Generally, they found that the critical length required for quenching is proportional to the turbulent flame velocity. Palmer suggested that a correlation of the form $L \sim S_t d_H^2$ would best represent his data on crimped ribbon and perforated blocks. Further work by Wilson and Attalah (1975) and Wilson and Flessner (1978) confirms this form of correlation, and they recommend that a proportionality factor of 2 sec-cm^{-2} be applied to the correlation, so that the final correlation for the minimum length is as given in Eq. (5-3).

TABLE 5-2
Equations for Calculation of Hydraulic Diameter for Various Flame Arresters

TYPE OF ARRAY OF OPENINGS	SKETCH	HYDRAULIC DIAMETER
TUBE BUNDLE OR PERFORATED PLATE	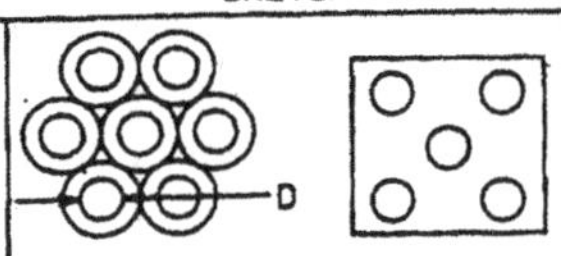	$D_H = D$
SCREEN OR GAUZE	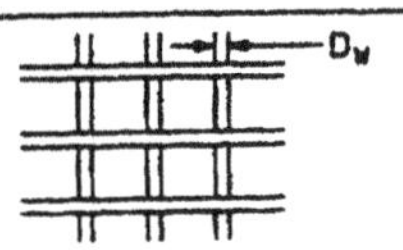	$D_H = 1/M - D_W$ M MESHES PER UNIT LENGTH
STACKED PLATES		$D_H = 1.4\ D_P$
CRIMPED RIBBON	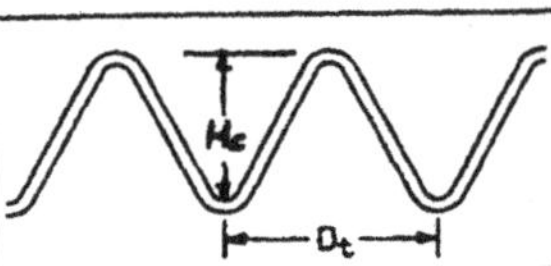	ISOSCELES TRIANGLE: $D_H = 0.83H_c$ EQUILATERAL TRIANGLE $D_H = 0.67H_c$
GRANULAR PACKED BED (SPHERES)		$D_H = 0.27\ D_s$
SINTERED METAL METAL FOAM OR WIRE PACK	MEAN PORE SIZE D_P 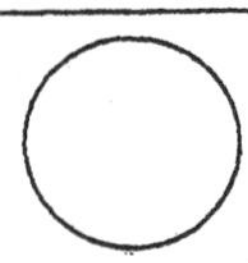	$D_H = D_P$

Correlations for Critical Diameter (Quenching Diameter)

Low-speed flames will only be quenched if the passageway diameter is below a certain critical value. This critical diameter can be calculated by the following equation (Mendoza et al. 1996):

$$N_{Pe,cr} = \text{Constant} = S_u(d_{cr})/\alpha \tag{5-5}$$

where $N_{Pe,cr}$ is the Peclet number for a critical passageway, S_u is the fundamental burning velocity of the gas and air mixture (m/sec), d_{cr} is the critical (quenching) diameter (m), and α is the gaseous diffusion coefficient of gas and air mixture (m^2/s). Values of S_u for various gases and vapors can be found in NFPA 68 (1998), and values for α can be found in Reid et al. (1977). Mendoza et al. (1996) recommend that a value of 65 be used for the Peclet number constant. Equation (5-5) allows computation of the critical (quenching) diameter of flame arrester passageways, based on physical properties of a flammable gas/vapor and air mixture. The equation is true for very slow flames because the quenching theory is based on laminar flow quenching only (Mendoza et al. 1996). Table 5-3 lists critical (quenching) diameters as circular apertures for various gases and vapors from experimental data (Smolensky 1999). Table 5-2 can be used to calculate the critical (quenching) diameters for noncircular apertures from the values listed in Table 5-3. See Chapter 9 for an example calculation.

Correlation for Quenching Length

Piotrowski (1991) recommends the following equation for the quenching length of a crimped ribbon flame arrester for high speed flames, based on the work of Wilson and Atallah (1975):

$$L = (S_t d_H^2)/100\nu \tag{5-6}$$

where L is the quenching length (cm), S_t is the turbulent flame speed (cm/sec), d_H is the hydraulic diameter (cm), and ν is the kinematic viscosity (cm^2/s) of the burned gas.

Correlations for Flame Speed

It has been shown by Palmer at the Fire Research Station (FRS) that a crucial variable governing the performance of a flame arrester is the flame speed incident on the arrester. The critical flame speed (minimum speed at which the flame can pass through the arrester) is discussed by Phillips and Pritchard (1986), drawing largely on the FRS work on propane–air mixtures at atmospheric pressure. A simple model based on heat removal from the flame yields the following relation for deflagration flame arresters:

$$V = (0.95nyp_o)/p \tag{5-7}$$

TABLE 5-3
Critical (Quenching) Diameters for Various Gases and Vapors (Smolensky 1999)

Gas or Vapor	Critical Diameter (mm)	Gas or Vapor	Critical Diameter (mm)
Acetic acid	5.6	Ethanol	3.0
Acetone	3.1	Ethyl acetate	3.1–3.4
Acetylene	0.7–0.85	Ethyl chloride	3.2
Amyl acetate	2.9	Ethyl ether	2.6
Benzene	2.4–3.2	Ethylene	1.5–1.75
Butane	3.4	Ethylene oxide	1.6
Iso-Butane	2.6	n-Heptane	2.5
1,3-Butadiene	1.5–2.1	n-Hexane	3.3
Isobutyl alcohol	2.9	Hexene	3.3
1-Butylene	2.2–3.0	Hydrogen	0.7–0.9
Isobutylene	3.5	Methane	2.8-3.7
Carbon disulfide	0.7	Methanol	2.2-2.7
Carbon monoxide	2.9	Methylcyclohexane	3.4
2-Chloropropane	4.3	Methylcyclopentane	3.6
Cyclohexane	3.4	2-Methylpentane	3.5
Cyclohexene	3.3	2-Methylpentene	3.5
Cyclopentane	3.3	2-Methylpropane	3.7
Cyclopentene	3.6	Iso-Octane	2-9–3.0
Cyclopropane	2.8	n-Pentane	2.4–3.4
n-Decane	2.4	Propane	2.9–3.0
Diethyl ether	2.2	Iso-Propanol	3.1
2,2-Dimethylbutane	3.6	Propylene oxide	1.7
2,3-Dimethylbutane	3.6	2,2,3-Trimethylbutane	3.6
2,3-Dimethylpentane	3.5	2,2,4-Trimethylpentane	3.2
3,3-Dimethylpentane	3.6	Vinyl acetate	2.8
2,2-Dimethylpropane	4.0	Vinyl chloride	2.7
Dimethyl sulfide	2.2	Xylene	3.1
Ethane	2.5		

where V is the flame speed at which the arrester fails (m/s), n is the number of apertures per unit area of arrester surface (cm^{-2}), y is the thickness (width) of the arrester elements (cm) (see Figure 5-1), p_o is atmospheric pressure (bar absolute), and p is the explosion pressure at the arrester (bar absolute). To obtain agreement with results obtained for an arrester mounted in a duct with a bend or obstruction, it was necessary to include the term p_o/p, which takes into account the increased explosion pressure. Another limitation of this equation is that it is only valid for crimped ribbon elements with apertures that are not more than half as wide as the quenching diameter of the gas mixture and with diameters equal to the test duct or pipe. Strictly speaking, it is only valid for propane–air flames since it was derived using the combustion properties of propane. However, it can be used for other hydrocarbon–air flames with similar combustion properties without modification.

Another empirical equation given by Phillips and Pritchard (1986) for the critical flame speed of crimped ribbon, wire gauze, and perforated plate arresters is as follows:

$$V = (0.38ay)/d^2 \quad (5\text{-}8)$$

where V is the flame speed at which the arrester fails (m/s), a is the fractional free area of the arrester surface, y is the thickness (width) of the arrester elements (cm), and d is the diameter of the apertures (cm). This

Notes to Table 5-3

1. The experimental values of the critical diameter were determined over a wide range of temperatures and pressures.

2. The measurements were made using different experimental apparatus such as narrow pipes, diaphragms with a round hole, flat slots, etc. Values of the critical diameter determined using noncircular apertures have been converted to equivalent circular apertures.

3. Data sources:

a. Holm, J. N., *Phil. Mag.*, 14, 18ff (1932); 15(3), 29ff (1933).
b. Friedman, R., *Proc. 3rd Symp. Combust.*, 110–120 (1949).
c. Friedman, R. and Johnston, W. C., *J. Chem, Phys.*, 20, 919–920 (1952).
d. Simon, D. M. et al., *Proc. 4th Symp. Combust.*, 126–138 (1953).
e. El'natanov, A. N. et al., *Fiz. Goreniya Vzryva*, 11(1), 142–144 (1975).
f. Belles, F.E. and O'Neal, C., *Proc. 6th Symp. Combust.*, 806ff (1956).
g. Potter, A. E. and Berlad, A. L., *J. Phys. Chem.*, 60, 97–101 (1956).
h. Wilson, C. A., *Ind. Eng. Chem.*, 51, 560ff (1959).
I. Grove, J. R., *Proc. 3rd IChemE Symp. Chem. Process Hazards Spec. Ref. Plant Des.*, 51–54 (1967).
j. Harris, M. E. et al., *Proc. 3rd Symp. Combust.*, 80–89 (1949).
k. Hieftje, G. M., *Appl. Spectrosc.*, 25(6), 653–659 (1971).
l. Strizhevskii, I. I. et al., *Powder Metallurgy (Kiev)*, 6, 88ff (1979).
m. Maekawa, M., *Combust. Sci. Technol.*, 11(3–4), 141–145 (1975).
n. Zel'dovich, Y. B., *J. Exp. Theor. Phys.*, 11(1), 159ff (1941).
o. Spalding, D. B., *Proc. Roy. Soc.*, A240, 83ff (1959).
p. Mayer, E., *Combust. and Flame*, 1, 438ff (1957)

equation is applicable to a deflagration, but not a detonation. The value of the flame speed, V, includes a safety factor that takes account of experimental error. This equation is derived from the work of Palmer, and therefore has the same limitations as Eq. (5-7). Equation (5-8) does not take into account the effect of explosion pressure and thus as given above will be valid only for low flame speeds where the pressure does not rise substantially from atmospheric, for example, in short lengths of straight pipe. If English units are used, V is in ft/s, y is in inches, d is in inches, and the constant is 0.50, rather than 0.38.

For noncircular apertures, the equivalent hydraulic diameter should be used. For crimped metal ribbon elements, the equivalent hydraulic diameter of a right isosceles triangle is 0.83 times the crimp height, and the thickness (width) should be at least 0.5 inches (HSE 1980).

For a wire gauze element, Eq. (5-8) applies only to a single layer of gauze, and the thickness, y, is twice the wire diameter. An increase in flame speed of about 20% of the original value of V may be obtained for each additional layer up to a maximum of five, but additional layers provide no further advantage (HSE 1980).

The *HSE Guide* (1980) presents two equations for detonation flame speeds, one for crimped metal ribbon arresters, and one for pipes.

For crimped metal ribbon arresters it provides the following equation for the maximum flame speed obtained by Cubbage (1959) for the stoppage of town gas–air mixture detonations by arresters with a crimp height of 0.017 inches:

$$y = 1.3(V^{1/5}) - 4.4 \tag{5-9}$$

where y is the metal element thickness (width) in inches, and V is the flame speed in ft/s.

Detonation flame speeds in pipes may be calculated approximately by the following equation:

$$V = 300[(\gamma^2 - 1)Q]^{0.5} \tag{5-10}$$

where V is the detonation velocity (ft/s), γ is the ratio of specific heats of the burned gases at the flame temperature, and Q is the heat of combustion of the gas mixture (cal/g). Detonation velocities depend primarily on the gas mixture composition, but not on the run-up length and only slightly on the gas pressure. Detonation velocities in small pipes are slightly slower than predicted by Eq. (5-10), because of the cooling to the walls, and this fact provides a margin of safety in the calculation.

As mentioned earlier, the detonation velocity depends on the composition of the gas mixture; for propane and other saturated hydrocarbons, and for many solvent vapors, a value of 5800 ft/s can be used for mixtures with air. The detonation velocity for town gas–air and hydrogen–air mixtures can be taken as 7000 ft/s. Detonation flame velocities can also be calculated by chemical equilibrium computer programs (Gordon and McBride 1976).

5.3.5. Burnback Resistance

There is always the possibility that a flame may stabilize on a flame arrester element and if not quenched in time, may heat the arrester to the degree that the flame may pass through it or even destroy it. Most of the flame arrester standards require endurance burn tests. However, endurance burn tests are presently subjective owing to the need to optimize variables as the test is underway. It is essential that burn tests be carried out in the orientation to be used in the actual service. For example, if the flame is stabilized on the upper face of a vertically mounted arrester, the arrester will receive minimal heat flux. This could lead to a significant overestimation of the arrester's endurance in actual service should the arrester encounter a flame on its lower face or be horizontally mounted (CCPS 1993).

The new CEN (2001) flame arrester standard gives details of the experimental set-up and specifications of gas–air mixtures to be used for short time and endurance burning tests for dry type arresters, hydraulic (liquid seal) arresters, and high velocity vent valves. Test gases include propane, ethylene, and hydrogen for short time burning tests (one minute maximum), and hexane, ethylene, and hydrogen for endurance burning tests (two hours minimum).

The US Coast Guard regulations also give details of endurance burning tests. They classify detonation arresters as either Type I or Type II. Type I arresters are acceptable for applications in which a stationary flame may rest on the device, and Type II arresters are suitable for applications in which stationary flames are unlikely to rest on the device. Other recommended safeguards to be provided to prevent flame passage when a stationary flame occurs, are a temperature sensor and automatic shutoff valve, or nitrogen or steam purge (see Section 5.3.18).

Endurance burn test procedures performed in accordance with UL and FMR are described in Chapter 8.

To predict the capability of a flame arrester to cool hot combustion gases, the U.S. Bureau of Mines has developed an equilibrium model and one- and three-dimensional transient thermal models of a flame arrester, which are used to predict the heat losses from the arrester and the maximum temperatures developed (Edwards 1991).

Capp (1992) has investigated the effects of propane–air mixture composition and flow rate on the temperature rise of a rigid element flame arrester in endurance burning. Different types of flame attachment with corresponding curves of temperature rise against time are presented. The mixture and flow rate giving the highest temperature increase are identified. Suggestions are made for the revision of the test procedures in the IMO, BSI, and USCG standards.

White (2000) points out that the type of fuel used may have a significant effect on endurance burn testing and endurance burn performance. The concern is that the test fuels commonly used (propane, ethylene, hydrogen, etc.) may not give good representation of their NEC gas group since they have higher AITs than other vapors and gases in their groups (i.e., gases or vapors with lower AITs might more readily reignite on the protected side of the arrester element). Therefore, tests for endurance burning should state specifically which gas or vapor was used for the test.

5.3.6. Pressure Drop Limitations

Flame arrester pressure drops must be taken into consideration when selecting and specifying them, especially when they are installed on vent systems of low pressure storage tanks, such as API-type tanks. If the total system pressure drop exceeds the design pressure of the tank, the tank will be overpressured and may be structurally damaged. Refer to Section 6.1 for additional discussion of this topic.

In many situations several types of flame arresters are suitable for the application. Often pressure drop during normal operation and its possible increase over the flame arrester's service life may be the primary determinant in selection of a suitable unit.

Data on the flow resistance of flame arresters are available from manufacturers' catalogues in the form of graphs. From the graphs, the flow resistance may be predicted in terms of the gas velocity, the model number or the aperture size of the arrester element, its thickness and porosity, etc. The predicted value is for a dry, clean arrester. Two additional factors affect pressure drop. These are element fouling and liquid accumulation (usually in horizontally installed arresters), which will depend on the application involved. Designing the arrester for a 20% fouling factor will in many instances allow for less frequent element removal and cleaning without operational problems. A larger element allows for a greater dispersal of particulates with a correspondingly lower pressure drop.

Depending on the design, the pressure drop through a crimped metal ribbon flame arrester may exceed that of a parallel plate type of the same size. If a crimped metal ribbon type arrester is used for end-of-line service on an API-type tank, it may be necessary to specify a larger size arrester than a parallel plate unit to achieve the desired low pressure drop.

Pressure taps are often specified on each side of an in-line flame arrester for checking the pressure differential across the arrester while in service. The taps and instrument installation must not provide a flame bypass around the flame arrester. By monitoring pressure drop, fouling or plugging may be detected early, if there is a sufficient vapor flow rate.

5.3.7. Fouling and Plugging Potential and Protection

As the majority of dry flame arresters contain narrow apertures, especially detonation arresters, they may easily become blocked or plugged in many ways. An end-of-line flame arrester can become plugged by accumulation of atmospheric debris such as dust, sand, fibers, chemicals, and other materials borne by the wind. There are reports of obstruction caused by mud-dauber wasp nests and bird nests, which can sometimes be minimized by screens. Pluggage has also occurred due to condensation and solidification of vapors from heated liquids in a vessel. This can be minimized by heating the arrester. Kletz (1982) describes a flame arrester for use with high melting point materials. Because of the increased temperature of the arrester, it is important to assure dependable operation in stopping flames at the elevated temperature.

In-line flame arresters, both deflagration and detonation types, can also become plugged for a variety of reasons. Sometimes rags, scale, and other construction debris can be found in the piping systems. In-line flame arresters, with their small apertures, serve as efficient strainers in flowing pipelines. Thus anything carried by the gas stream can potentially be caught by the arrester. Vessels to which solids are added, or in which solids are mixed with liquids, often have solids entrained with the vent vapors, which can plug flame arresters. Even very small particles can sometimes stick on the large surface area of an arrester and gradually accumulate to form large masses. Through the above and possibly additional mechanisms, an in-line flame arrester can become plugged more readily than an end-of-line flame arrester. Thus, prevention of plugging of in-line arresters can be very difficult, and if there is a strong potential for plugging, dry in-line flame arresters should be avoided. Hydraulic flame arresters, or other combustion safeguarding methods described in Chapter 3, should be considered.

Flame arrester design should allow for mitigation of condensate build-up that restricts gas flow. The potential for condensate accumulation is greater for horizontal flame arrester installations than for many vertically installed flame arresters whose elements are more readily self-draining. In some cases liquid can obstruct a significant portion of the free-flow area creating a large pressure drop and additional problems should freezing or corrosion occur. Allowance for condensate accumulation should be made for periods when automatic or manual drainage is not possible. Any drains provided should neither provide a flame path around the element nor provide leakage in either direction when closed.

As noted in Section 5.3.6, if a flame arrester mounted on a breather vent nozzle of an atmospheric tank becomes plugged, overpressure may occur when the tank is filled or a vacuum may occur when the tank is emptied. Overpressure or vacuum may also be caused by ambient temperature change. See Section 6.1 for additional information.

One of the more effective preventive measures is flame arrester monitoring, which will be discussed in Section 5.3.18.

5.3.8. Unwanted Phases

Unwanted phases such as liquid slugs, tramp metal, etc., can damage dry flame arrester elements so that they have to be replaced, or they can also block or plug the arrester free-area. It is not always obvious when an arrester will be impacted by a liquid slug or tramp metal. However, when this is suspected or has already occurred in a process, several things can be done to avoid the problem as follows:

- provide a knock-out device or debris trap (such as is used for frangible graphite rupture disks) ahead of the flame arrester
- use a hydraulic (liquid seal) flame arrester rather than a dry type flame arrester

5.3.9. Material Selection Requirements

When the materials handled are noncorrosive, the flame arrester vendor's standard materials of construction for the housing and arrester element are commonly used. For noncorrosive service housings are normally available in aluminum, carbon steel, ductile iron, and 316 stainless steel; the elements are commonly available in aluminum or 316 stainless steel.

In some cases, it may be desirable to specify an all stainless steel arrester to avoid contamination of the stored liquid by rust or other particles from the housing. Note that certain metals may catalyze polymerization of vapors in gas streams containing monomers, which will then be deposited on the flame arrester elements and hasten blockage. Also note that rust may be transported to the flame arrester from upstream sources.

Consideration should be given to the corrosion allowance for flame arrester elements and housings. Element corrosion will reduce the effectiveness of the arrester to quench a flame and withstand the destructive pressure effects of a detonation. Also, increased element thickness will allow for greater periods between element replacement.

For corrosive service, such as processes involving acid components, dry flame arresters are available in higher alloys such as tantalum, Hastelloy B and Hastelloy C construction for both the housing and element. In Europe, one manufacturer can provide flame arresters with fluoropolymer thermoplastic solid arresting elements and lined housings up to a nominal size of 4 inches, which have been successfully used in the chemical industry (Halstrick and Kirchner 1996). Because of the high cost for tantalum, Hastelloy B or Hastelloy C dry-type flame arresters, it may be more economical to use a proprietary carbon steel hydraulic arrester with a suitable polymeric lining.

Corrosion resistance can also be provided by lining a carbon steel housing with stainless steel or other alloy.

5.3.10. Special Design Options

Several special design options are available and have been provided by at least one U.S. flame arrester manufacturer as follows (Mendoza 1999):

- a water-cooled flame arrester with cooling tubes drilled through the arrester element for a high-temperature application
- an air-cooled flame arrester with nozzles welded to the arrester housing for introduction of cooling air to cool the flame arrester element for a gas (ethylene oxide) thermal oxidizer feed nozzle
- flame arresters with jackets welded to the housing for heating of the arrester with steam, hot water, or heat transfer fluids for high melting point materials

Nozzles can be installed on the flame arrester housing for injecting snuffing steam, carbon dioxide, or a suppressant into the arrester to extinguish a standing flame on the arrester element.

Nozzles can also be installed on the flame arrester housing for the injection of hot cleaning solution for on-line cleaning of polymeric deposits (Dafft 1999). This is especially useful for very large flame arresters that cannot be easily removed from service.

5.3.11. System Constraints

The length and configuration of the piping system on the run-up (unprotected) side of the arrester can determine whether a detonation will occur.

This includes the positioning of any turbulence-promoting flow obstructions such as tees, elbows, and valves, which can significantly increase the flame speed. Thus, it is of great importance to establish where ignition might occur in the system and how this will affect the flame path to the arrester. Pipe diameter also affects the distance required for deflagration to detonation transition (DDT); larger pipe diameters typically require longer run-up distances for detonation. Testing has shown that reductions in pipe diameter along the pipe run dramatically increase the flame speed and pressure. The opposite effect occurs with increased pipe diameter. See Section 5.3.15 for a discussion of the effect of changes in pipe diameter. Although detonations may fail on encountering branches into smaller pipe diameters, run-up to detonation may reoccur (Frobese and Forster 1992).

5.3.12. Mixture Composition

The ranges of mixture composition plus operating temperature and pressure are important. Unless a detonation flame arrester is used, it is essential to ensure that DDT will not occur after a flame enters the system. Even if a detonation flame arrester is used, it is important to use the correct type of arrester for the service. For example, if it is possible for a flame to approach from either direction, a bidirectional flame arrester must be selected.

Detonations can only take place within the "detonable limits," which fall inside the flammable limits. Mixture composition may vary widely in some systems, and this can further depend on the number of connections feeding a manifolded header. There are few cases where it can be assumed that a nondetonating flammable mixture will be produced; hence, most in-line flame arrester applications involve the use of detonation flame arresters.

Testing protocols presently do not address systematic variation of the mixture composition for all types of flame propagation. However, it is possible that worst cases may exist away from the near-stoichiometric mixtures used, particularly for restricted-end deflagration flames. Restricted-end (using an orifice) deflagration tests create additional pressurization in the flame arrester due to increased flow resistance plus reflected shocks created by the orifice. Such conditions may cause the flame arrester to fail via flame penetration through the element. The failure mechanism has not been definitely established but one likely factor is the decrease of quenching diameter with increased pressure, and another is an elevated burned gas exit temperature. The temperature of burned gas exiting the flame arrester element will depend on various factors, including the penetration distance prior to flame quenching. This penetration distance increases as pressure increases, and consequently, quenching diameter decreases. A likely failure mode (short of complete flame penetration through the element) is that the burned gas exiting the flame arrester is sufficiently hot to ignite unburned gas on the protected side. This failure mode defines the borderline failure case.

For most hydrocarbons in air the most sensitive composition for such parameters as autoignition temperature, burning velocity, and minimum ignition energy is on the rich side of the stoichiometric composition, often corresponding to a stoichiometric ratio of about 1.4. The likelihood of unburned gas ignition on the protected side of the flame arrester due to hot gas breakthrough should be sensitive to its composition. Therefore, restricted-end deflagration testing should also address mixtures that are known to give the lowest MESG values or are otherwise considered most sensitive to ignition, for example, at about 1.4 times the stoichiometric composition.

5.3.13. Operating Temperature and Pressure

Elevated temperature or pressure affects flame arrester performance. With respect to changes in pressure, the process of flame quenching becomes more difficult as the pressure increases, even though flame speeds may not be much affected. The difficulty arises because more heat is released per unit volume of gas mixture as the pressure is raised, which in turn decreases the required quenching diameter and the autoignition temperature. The increase in the volumetric heat release is proportional to the increase in the absolute pressure. The effect of changes in gas temperature is more complex. The fundamental burning velocity increases with temperature, but the density of the gas is reduced, which causes the heat released per unit volume of gas to diminish. The quenching diameter is found to be approximately inversely proportional to the square root of the absolute temperature; that is, the quenching diameter decreases as the temperature rises (HSE 1980).

Testing should address the maximum operating temperature and pressure normally seen at the arrester location. This does not include certain pressure upsets (such as emergency shutdown) that produce unusually high system pressures. In many cases it may not be possible to design an arrester that will function effectively during upset conditions, and other protective measures should be considered (e.g., venting, suppression).

It may be necessary to position flame arresters away from heat sources that could cause the allowable operating temperature of the arrester to be exceeded. Positioning must be made with due consideration of DDT constraints. See Table 5-4, which shows the relationship of run-up distance to

TABLE 5-4
Effect of Run-up Length on Propagation Velocity and Overpressure (CCPS 1993)

Run-up Length (ft)	Propagation Velocity (ft/s)	Detonation	Overpressure (psi)
1	15	No	6.9
6	250	No	16.1
19	400	No	48.3
24	7360	Yes	2044

Note: The above data are from tests in a straight 3-inch diameter pipe using a test gas consisting of 4.3 volume percent propane mixture in air, initially at 23 psia.

propagation velocity and overpressure. These data are for tests in a straight 3-inch pipe, for a 4.3 volume percent propane mixture in air initially at 23 psia (CCPS 1993). Data for other gases show a similar relationship: As run-up distance increases, propagation velocity, overpressure, and the potential for detonation to occur increases. Testing should allow the presence of the heat source (including connection to boilers or incinerators) to be simulated. Alternatively, the flame arrester can be cooled (see Section 5.3.10).

Operating pressure is extremely critical. The maximum allowable operating pressure may be as small as a few psig depending on the gas involved. This constraint must be ascertained when selecting both the type and location of the flame arrester. This is true for both deflagrations and detonations. Note that the operating pressure relevant to flame arrester performance is the initial pressure on the unprotected (flame) side. Thus, it is important to consider the pressure drop across the flame arrester including the effects of partial blockage due to fouling.

5.3.14. Ignition Location

Where a single tank vent system employs an end-of-line deflagration flame arrester, only flames entering from outside the system need be considered. In selecting a deflagration flame arrester, it is necessary to (1) ensure that DDT cannot occur between the atmospheric ignition source (for example, an external flame or lightning strike) and the arrester, and (2) test the arrester under the configuration and conditions to be encountered in the actual installation.

Although the types of plant and equipment using flame arresters vary widely, they often have some common features. Frequently the system is not completely enclosed, and it is usual for a part to be open to the atmosphere or to a relatively large reservoir (tank or vessel) either through a duct opening, or a restricted opening such as a nozzle. Thus if an explosion occurs, there is a preferential direction for the gas to move as soon as the pressure begins to increase. This preference in gas movement direction affects the performance required of the arrester. Figures 5-12 and 5-13 illustrate a simplified system showing the effect of ignition source location for a nonflowing and flowing system, respectively (HSE 1980). The behavior of more complex systems can often be explained using this simplified illustration.

Nonflowing System

The system (see Figure 5-12) initially contains a flammable gas–air mixture at atmospheric pressure. The mixture is initially stationary. Three cases are described as follows:

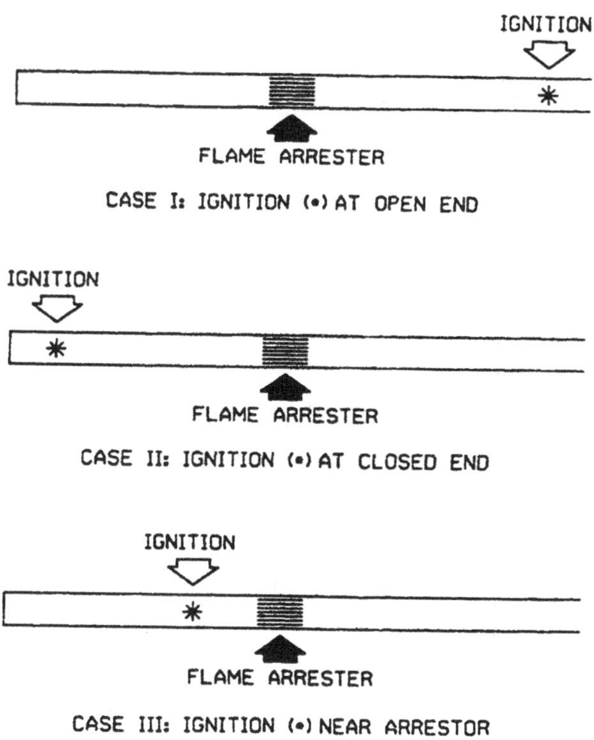

FIGURE 5-12. Effect of ignition source location—nonflowing system.

CASE I
Ignition occurs at the open end of the duct. The flame propagates into the duct until it reaches the flame arrester element where it is quenched. In this case, the amount of heat that must be dissipated by the arrester is relatively small because the hot combustion gases discharge through the open end of the duct.

CASE II
Ignition occurs at the closed end of the duct. The unburned gases are displaced through the flame arrester ahead of the flame. The flame accelerates through the unburned gases. The unburned gases are displaced through the flame arrester as the burned gases expand. The flame is quenched at the element, and the residual hot combustion gases gradually cool by heat loss at the duct walls.

CASE III
Ignition occurs near the flame arrester in the closed end of the duct, and the flame propagates in two directions. Soon after ignition, a slowly moving flame arrives at the flame arrester; meanwhile, another flame propagates toward the closed end of the duct, and the hot combustion gases are expelled through the flame arrester. In this case, the arrester must be massive enough to quench the near flame and also be sufficiently strong to withstand the pressure due to the moving gas.

If in the above cases the gases were flowing continuously toward the open end of the duct, the flame would behave differently, as discussed below.

Flowing System
In this system (see Figure 5-13) a flammable gas–air mixture at atmospheric pressure is introduced continuously.

CASE I
In this case ignition occurs on the downstream side of the flame arrester. If the velocity of the gas is less than the flame speed, the flame stabilizes on the flame arrester and continues to heat it until the gas flow is stopped or the flame is quenched by other means. Continued heating could cause ignition of the gas on the opposite side of the flame arrester.

If the velocity of the gas is sufficiently greater than the flame speed, the flame is swept out of the duct, or it stabilizes at the exit of the duct.

CASE II AND III
In these cases, it is assumed that the flame stabilizes on the upstream side of the flame arrester and hot gas flows through the arrester, heats it, and

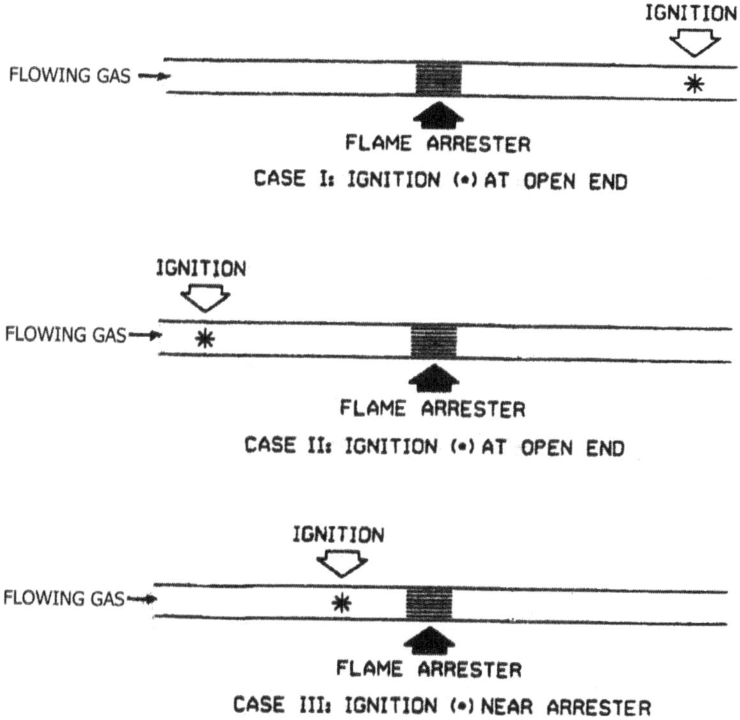

FIGURE 5-13. Effect of ignition source location—flowing system.

allows the flame to pass through. In either case, the flame arrester fails, and a flow of hot combustion gases passes through the flame arrester. This scenario is for a gas with a reasonably high velocity so that flashback cannot occur upstream. At lower gas velocities the flame will flash back upstream, and the flame arrester will not receive an on-going source of unburned gas.

It is emphasized that the system shown in Figure 5-13 represents only a simplification of actual plant installations, which may be more complex. If it is not obvious at which point ignition is likely to occur, a flame arrester installed in an actual plant may have to be selected to face a combination of the conditions shown in Figure 5-13. Therefore, for manifolded vent systems, the arrester should be a bidirectional, detonation type, and both sides of the arrester element should be provided with thermocouples to detect a stable flame.

In Cases I and II, depending on the run-up distance and gas velocity, there is a greater possibility of DDT than in Case III since the ignition source is further away from the flame arrester.

Lapp and Werneburg (2000) present test data showing that ignitions originating in tanks or vessels (called pre-volume vessels) connected to piping systems propagate as high pressure deflagrations and detonation flame fronts. The tests were conducted using a 6.5 volume percent ethylene/air mixture, a 4-inch line, a pre-volume vessel approximately 0.85 cubic meters (30 cubic feet in volume), and the separation distance between the flame arrester and the pre-volume vessel was 4.64 meters (15.22 feet). The ignition source was located both at the opposite end of the vessel from the discharge nozzle and close to the discharge nozzle. In all the tests with the pre-volume vessel, the flame arrester failed. This same arrester stopped all flame fronts in a similar straight pipe test without a pre-volume vessel attached, and at an initial pressure of 34 kPa (5 psig). They state that both end-of-line and in-line deflagration flame arresters will fail. The pre-volume and run-up distance will duplicate the "larger line size" phenomenon and will cause all flame arresters (end-of-line, in-line deflagration, and detonation flame arresters) to fail. They also recommend that when pre-volume vessels are present, only detonation flame arresters should be used, and that the arrester location must be 120 pipe diameters from the vessel unless pre-volume/piping ratio evaluations are done on the actual vessel/pipe configuration in a test. However, other flame arrester manufacturers do not agree with these recommendations, and accept the location of a flame arrester close to or directly on the vent nozzle of the pre-volume vessel.

5.3.15. Changes in Pipe Diameter

Flame arrester performance can be impaired if the pipe diameter increases within a minimum distance from the arrester in the flame approach direction. A study by Lapp and Vickers (1992) shows a marked decrease in maximum allowable operating pressure when the pipe size increases within 120 pipe diameters of the arrester. The configuration considered was in-line using a transition piece between the two pipe sizes. The testing showed that when a USCG-accepted detonation flame arrester was connected to a larger line size, it failed to stop low pressure and low speed flame fronts in operating conditions far less severe than original test conditions. All deflagration arresting capability was also lost and the flame arrester failed (Lapp and Werneburg 1999).

Figure 5-14 shows the required installation configuration when a larger pipe is located upstream of a flame arrester (i.e., it is necessary to install a pipe at least 120 pipe diameters long of the same size as the flame arrester inlet nozzle between the larger pipe and the flame arrester).

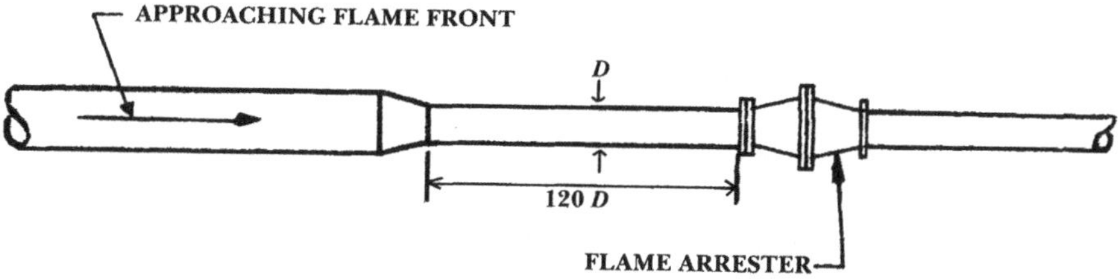

FIGURE 5-1. Required piping installation for changes in pipe diameter upstream of a flame arrester.

At first glance, it would appear that the results by Lapp and Vickers (1992) do not agree with the findings of Frobese and Forster (1992) since the recommended separation distance of Lapp and Vickers is far in excess of that proposed by Frobese and Forster.

The latter investigation of large/small diameter pipework junctions concluded that a "detonation arresting device will prevent flame transmission if it is installed at a distance up to a maximum of 20 pipe diameters from the branch point." However, there are two variances in the testing arrangements that may explain the contrast in the two required separations. Lapp and Vickers conducted testing on a straight length of pipe; however, the Frobese and Forster tests incorporated a tee piece, with the larger line open to the atmosphere. Hence, detonation propagation into the smaller pipe was side-on and much of the available energy from the detonation vented to the atmosphere rather than being passed through the flame arrester in the branch line.

An additional finding by Lapp and Vickers (1992) is that a change in pipe diameter (as above) has the largest detrimental effect during restricted-end deflagration testing.

5.3.16. Location and Orientation

All flame arresters should be located so that they are readily accessible for maintenance. If an end-of-line flame arrester is on a vent pipe from an indoor vessel, the pipe should be routed above the building roof so that flammable vapors are not discharged into the building. Also, the flame arrester should be installed so that the discharge opening is at a minimum height above the roof so that personnel on the roof will not be directly impacted by the vent vapors. Some companies specify the minimum height to be seven to ten feet above the roof.

Flame arresters on outdoor reactors, storage tanks, and process vessels may be of the in-line, deflagration type with a short vent pipe connected in such a manner so that hot vented vapors do not impinge on the vessel head. This avoids overheating the head by flame impingement should the vent become ignited.

With respect to detonation flame arresters in vent manifold systems, they can be located at any distance along the manifold as long as they are easily accessible for maintenance..

Where possible, all flame arresters should be installed so that the elements will be self-draining. If liquid can accumulate in the housing, provision should be made to drain any accumulated liquid that could block the flow of gases through the flame arrester.

See Section 6.4 for more discussion of this topic.

5.3.17. Reliability

In order to ensure the reliability of a flame arrester (ability to quench a propagating flame or withstand a stationary flame) a number of factors must be taken into consideration as follows:

- Dry type flame arresters with small apertures, especially detonation arresters, should not be used in services where dust is present as the arresters will probably plug.
- Flame arresters should only be used for groups of gases and vapors for which they have been certified by testing. Testing should be done according to an accepted standard.
- A flame arrester should not be installed in pipes larger than that for which the arrester has been tested.
- Monitoring the pressure drop of a dry type flame arrester is advisable to determine if the arrester is plugging.
- Temperature monitoring of a flame arrester to detect a stabilized flame (endurance burning) should be provided if it is expected that this may occur, or if such an event has already occurred in the past.
- Regularly scheduled maintenance should be instituted.
- Operating pressure must be in the range for which the flame arrester has been tested.
- Location and piping configuration must consider the possibility of DDT. For example, for end-of-line and in-line deflagration flame arresters, the location must be identical to the test piping configuration or a DDT may occur.

More detailed discussion of the above considerations is presented in other sections of this book.

5.3.18. Monitoring and Instrumentation

Proper monitoring and instrumentation of flame arresters are vital to ensure the ability of arresters to function reliably. This is especially true for in-line, detonation arresters, which have very small apertures and thus a greater potential for plugging or fouling. Pressure taps should be provided on each side of an in-line flame arrester for checking of the pressure differential across the arrester while the gas in the pipeline is flowing. By keeping records of pressure drop, the early occurrence of fouling or plugging can be detected. It is important that the differential pressure instrument does not provide a flame path around the arrester.

Temperature sensors are often provided on the unprotected side(s) of the arrester element and as close as possible to the respective faces of the element. This is especially needed when the flame arrester is at the end of a manifold near the entry nozzle of a flare or thermal oxidizer where flame impingement could occur. The purpose is to sense a stabilized flame at the flame arrester element quickly and to send a signal to alert operating personnel to this problem or to cause automatic action, such as fast valve closure, to take place. Since fast response may be required of the sensor, it should not be inserted in a thermowell. For example, if the sensor is to be a metal-sheathed thermocouple (which is the normally preferred temperature measuring instrument), it must be of small diameter and located at the top of the flame arrester or pipe to ensure quick response. The alarm or automatic corrective action should be initiated from a relatively low temperature rise to speed up response action rather than waiting for the sensor's temperature to approach that of the flame. The temperature sensor(s) can be interlocked to a variety of mitigating actions such as shutting a fast-acting valve in the piping, admitting a large flow of nitrogen, steam, or carbon dioxide, or activating an explosion suppression system.

For hydraulic (liquid seal) flame arresters, the following safety features and instrumentation are recommended:

- a local level indicator with an optical display to indicate liquid seal height
- control room level indication in addition to the local level indicator
- a level controller and control valve in the makeup seal liquid line to automatically and rapidly admit liquid when low level is reached
- a temperature sensor in the seal liquid for freeze protection
- a temperature sensor above the seal liquid to indicate a stabilized flame
- provisions to protect the hydraulic flame arrester upon detection of high temperature above the seal liquid, e.g., which will admit a large quantity of snuffing gas such as steam or nitrogen into the arrester

- freeze protection if water is the seal liquid and the hydraulic flame arrester is located outdoors where there is a potential for freezing, or venting occurs at low temperature (vent gases entering the hydraulic flame arrester are below 32°F)
- reliable or back-up seal liquid supply
- continuous overflow and makeup to keep from building up organics in the seal liquid (usually at the surface of the seal liquid)

In some cases periodic analysis of the seal fluid should be done to check for the accumulation of fouling materials, such as gels, which may be due to reaction of the seal fluid with process gases and materials.

5.3.19. Inspection and Maintenance Requirements

Regularly scheduled maintenance is a "must" to ensure that flame arresters are in operable condition when they are called upon to function. Normally, maintenance is performed during a turnaround or shutdown for major maintenance work. Flame arrester maintenance is often neglected for two reasons: (a) arresters are often located in some rather remote area of a plant (e.g., a plant tank farm), and (b) arresters are often thought of as not being essential to the operation of a process and are therefore not given adequate attention. Both of these factors can make flame arresters easy to disregard, even though they may be listed on a computer maintenance schedule. It is, therefore, imperative that flame arresters be given the same inspection and maintenance priority as other safety devices (safety valves, rupture disks, etc.).

Deflagration and detonation flame arresters should be inspected annually until operating experience indicates otherwise. Also, the need for frequent inspection and maintenance may affect the selection of one type of flame arrester over another type for a specific application. For example, a hydraulic (liquid seal) flame arrester may be more suitable than a dry, fixed-element, flame arrester if the latter requires frequent inspection and maintenance because of persistent plugging problems.

See Chapter 7 for a more detailed discussion of this subject.

5.4. Special Applications

This section discusses the selection and application of flame arresters for quenching flames from gases with high fundamental burning velocities and/or the propensity to self-decompose.

5.4.1. Hydrogen

Several types of flame arresters have been tested for hydrogen service and found acceptable for quenching of hydrogen-air and hydrogen–methane–air mixtures.

Howard et al. (1975) conducted experiments on three types of flame arresters for quenching fuel mixtures of 90 mole percent hydrogen and 10 mole percent methane with air. In the tests, the hydrogen–methane mixture could be mixed with air in any ratio. For most of the testing, the flammable mixture was held at known values in the range of stoichiometric to 10% fuel rich relative to stoichiometric. The three flame arrester types were a proprietary hydraulic arrester, a crimped metal ribbon arrester, and a velocity flame stopper. Tests were run at pressures of 4 and 12 psig and feed gas temperatures of ambient, 300°F, 400°F, and 500°F. In these experiments only the velocity flame stopper was able to stop all flame propagation, if there was a dependable, minimum flow rate of gas through the orifices. The velocity flame stopper was in the shape of a tee with 0.161-inch diameter holes.

Some crimped metal ribbon flame arresters have been tested for hydrogen service and can be used. IMI Amal had experiments conducted at the Fire Research Station on an end-of-line deflagration flame arrester and an in-line detonation flame arrester using hydrogen-air mixtures. Both arresters were crimped metal ribbon types, and were successful in stopping the flames. The end-of-line flame arrester was a 2-inch unit fitted with a stainless steel cowl, and was tested with a mixture comprised of 42 volume percent hydrogen and 58 volume percent air (FRS 1985). The in-line detonation flame arrester tests were done on a 2-inch unit, using a mixture of 37.5 volume percent hydrogen and 62.5 volume percent air (FRS 1990). Later tests were successfully performed on 6-inch units. Testing has also been done by Protectoseal at Southwest Research Institute on 2-inch and 3-inch end-of-line crimped metal ribbon units certified to meet the ASTM F1273 standard. Protego™ has both deflagration and detonation flame arresters, ranging in size from 10 mm to 400 mm, approved in Germany for mixtures of hydrogen and air in all ranges of concentration (1993). The housings for both the deflagration and detonation flame arresters made by Protego™ are hydrostatically tested to 10 bar (about 145 psig).

Some designs of hydraulic (liquid seal) flame arresters have been successfully tested for hydrogen service. NAO has designed and successfully tested and provided a hydraulic flame arrester for hydrogen–air applications (Straitz 1999). This design is for detonations and has dual liquid seal chambers with shock wave breakers. Rao (1980) also provides information on a hydraulic (liquid seal) flame arrester that was designed and used successfully for hydrogen service in a nuclear power plant.

5.4.2. Acetylene

Acetylene may propagate decomposition flames in the absence of any oxidant above certain minimum conditions of pressure, temperature, and pipe diameter. Acetylene, unlike most other gases, can decompose in a detonative manner. Among the different types of flame arresters that have proven successful in stopping acetylene decomposition flames are hydraulic (liquid seal) flame arresters, packed beds, sintered metal, and metallic balls (metal shot).

Sutherland and Wegert (1972) describe the successful use of the Linde hydraulic valve arrester in stopping an acetylene decomposition detonation. As previously noted, these flame arresters are no longer being made by Linde (now Praxair Inc,), but are still available from ESAB Welding & Cutting Products of Florence, SC.

Schwartz (2000) reports the successful use of the John Zink "bubble-screen" hydraulic flame arrester in acetylene service.

Packed columns, with both dry and wetted packings, have been used for many years in low pressure (up to 15 psig) acetylene service. They are usually packed with 1-inch Raschig rings. Design information about them is found in CGA pamphlet G1.3 (1970). Schmidt (1971) also discusses their design and use for acetylene service. As mentioned in Section 5.2.11, the packing size should be decreased as the operating pressure increases. However, Britton (2000) points out that experimental work carried out in Germany indicates that soot deposits on the packing improves the ability of a packed bed flame arrester to stop acetylene decomposition flames. This finding is very important with respect to the design of dry packed bed acetylene flame arresters. Unless soot deposits are removed between tests the results will be nonconservative when applied to clean in-service beds. Published design and test data must be carefully examined to determine whether this effect was recognized.

For high pressure (15–400 psig), two types of acetylene flame arresters were developed by Linde and are still available from Praxair on special order. The first type is available in 1.5-inch diameter by 3- to 4-foot-long cylinders packed with sintered metal, fine wire wool, or finely divided alumina. The second type is available in 6-inch diameter by 15-inch-long cylinders packed with round nickel shot.

Flame arresters for acetylene–air mixtures with a flame-arresting element of sintered metal are available from Western Enterprises of Westlake,

OH and Rexarc of West Alexandria, OH. These are small diameter units used mostly for oxyacetylene welding equipment.

Protego™ offers a crimped metal ribbon flame arrester approved in Germany for acetylene service. It is similar in design to their flame arresters for hydrogen service but the hydraulic diameter of the flame arrester apertures for quenching acetylene flames is 0.15 mm rather than 0.20 mm for hydrogen.

5.4.3. Ethylene Oxide

Ethylene oxide (EO) can also propagate decomposition flames in the absence of any oxidant under specific conditions of pressure, temperature, and pipe diameter. However, unlike acetylene, EO is not known to detonate in plant piping in the absence of an oxidant. This is based on operating experience. However, a recent study (Thibault et al. 2000) reports that DDT may occur in large diameter pipe (>12-inch diameter) under certain conditions. Deflagration flame arresters consisting of tube bundles of specified length and diameter have been used for many years in process units (Britton 1990). Recklinghausen (1978) of Chemische Werke Huls AG describes an alternative ethylene oxide flame arrester of the packed bed type. Schwartz (1999) also reports that the John Zink Bubble-Screen liquid seal flame arrester has been used successfully for ethylene oxide service.

5.5. Information That Should Be Provided to Manufacturers

To obtain the proper quotation from flame arrester manufacturers it is imperative that they be provided with all the information needed to select and quote the most appropriate arrester for the specific application.

Information that should be provided includes the following:

- type of arresters under consideration (end-of-line, in-line, hydraulic, etc.)
- gas or vapor being handled, including compositions during upsets, start-up, shutdown, etc., noting presence of corrosive or fouling components
- normal and maximum gas/vapor flow rate
- normal and maximum operating pressure
- normal and maximum operating temperature
- allowable pressure drop
- materials of construction of housing and arrester element
- optional fittings for drains, pressure taps, temperature sensors, etc.

- piping drawing showing all straight run lengths, elbows, tees, changes of direction, etc., if a flame arrester is in a manifolded vent system

Appendix A shows a sample flame arrester specification sheet for fixed element arresters which may be used to obtain quotations from manufacturers..

5.6. References

5.6.1. Regulations, Codes of Practice, and Industry Standards

API RP 2210. 2000. *Flame Arresters for Vents of Tanks StoringPetroleum Products.* 3rd ed. American Petroleum Institute, Washington, DC.

API RP 521. 1997. *Guide for Pressure-Relieving and Depressuring Systems.* 4th ed. American Petroleum Institute, Washington, DC.

ASTM F-1273-91. 1991. *Standard Specification for Tank Vent Flame Arresters.* American Society for Testing and Materials, West Conshohocken, PA.

CEN (Comité Européean de Normalisation) Std. EN12874. 2001. *Flame Arresters-Specifications, Operational Requirements and Test Procedures.* European Committee for Standardization, Brussels, Belgium.

CGA (Compressed Gas Association, Inc.). 1970 (Reaffirmed 1984). *Acetylene Transmission for Chemical Synthesis (Recommended Minimum Safe Practices for Piping Systems).* Publication G-1.3. Compressed Gas Association, Inc., Arlington, VA.

HSE (Health and Safety Executive). 1980. *Guide for Flame Arresters and Explosion Reliefs.* Booklet HS(G)11. Health and Safety Executive, Her Majesty's Stationery Office, London, England.

IEC (International Electrotechnical Commission) IEC 79-1A. 1982. First supplement to Publication 79-1 (1971), Electrical Equipment for Explosive Gas Atmospheres, Part 1, Appendix D, 2nd impression. International Electrotechnical Commission, 1 rue de Varembe, Geneva, Switzerland.

NFPA 30. 2000. *Flammable and Combustible Liquids Code.* National Fire Protection Association, Quincy, MA.

NFPA 325. 1994. *Fire Hazard Properties of Flammable Liquids, Gases, and Volatile Solids.* National Fire Protection Association, Quincy, MA.

NFPA 497. 1997. *Recommended Practice for the Classification of Flammable Liquids, Gases, or Vapors and of Hazardous (Classified) Locations for Electrical Installations in Chemical Process Areas.* National Fire Protection Association, Quincy, MA.

29 CFR 1910.106. 1969. Flammable and Combustible Liquids. Code of Federal Regulations. Washington, DC.

33 CFR Part 154, Subpart E—*Vapor Control Systems.* U.S. Department of Transportation, Coast Guard. June 21, 1990.

UL 525. 1994. *Standard for Flame Arresters.* 6th ed. Underwriters Laboratories, Inc., Northbrook, IL.

5.6.2. Specific References

Bjorklund, R. A. and Ryason, R. R. 1980. Detonation-Flame Arrester Devices for Gasoline Cargo Vapor Recovery Systems. JPL Publication 80-18. Jet Propulsion Laboratory, Pasadena, CA

Bjorklund, R. A., Kushida, R. O., and Flessner, M. F. 1982. Experimental Evaluation of Flashback Flame Arresters. *Plant/Operations Progress,* 1(4), 254–262.

Borger, G. G., Huning, W., Nentwig, H. U., Schweitzer, M., and Schellmann, E. 1991. Hydraulic Flash-Back Protection. *Proc. 3rd Intl. Symposium on Loss Prevention and Safety Promotion in the Process Industries,* 2, 1261ff.

Briesch, E. M. 2000. NEC Group Classification of Mixtures. Paper 5d, AIChE 34th Annual Loss Prevention Symposium, March 6–8, 2000, Atlanta, GA.

Britton, L. G. 1990. Thermal Stability and Deflagration of Ethylene Oxide. *Plant/Operations Progress,* 9(2).

Britton, L. G. 1996. Operating Atmospheric Vent Collection Headers Using Methane Gas Enrichment. *Process Safety Progress,* 15(4), 194–212.

Britton, L. G. 2000a. Using Maximum Experimental Safe Gap to Select Flame Arresters. *Process Safety Progress,* 19(3), 140–145.

Britton, L. G. 2000b. Personal communication from L. G. Britton, Union Carbide Corporation, to S. S. Grossel (March 1, 2000).

Broschka, G. I., Ginsburgh, I., Mancini, R. A., and Will, R. G. 1983. A Study of Flame Arresters in Piping Systems. *Plant/Operations Progress,* 2(1), 5–12.

Capp, B. 1992. Temperature Rise of a Rigid Element Flame Arrester in Endurance Burning with Propane. *J. Loss Prev. Process Ind.,* 5(4), 215–218.

Capp, B. and Seebold, J. 1991. Detonation Experiments in an 18-inch Pipe. Paper presented at the 1991 Annual AIChE Meeting session of Fundamental Chemical and Physical Processes in Combustion and Incineration I. Los Angeles, CA., November 17–22. American Institute of Chemical Engineers, New York, NY.

CCPS. 1993. *Guidelines for Engineering Design for Process Safety.* Center for Chemical Process Safety, American Institute of Chemical Engineers, New York, NY.

Cubbage, P. A. 1959. Flame Traps for Use with Town Gas/Air Mixtures. *Gas Council Research Communication GC 63.* The Gas Council, London, England.

Cubbage, P. A. 1963. The Protection by Flame Traps of Pipes Conveying Combustible Mixtures. *IChemE Symposium Series No. 15,* pp. 29–34. Institution of Chemical Engineers, Rugby, England.

Dafft, C. A. 1999. Personal communication from Tony Dafft, Rohm and Haas Company, to S. S. Grossel (September 1999).

Dickerman, D. 1999. Personal communication from Dwight Dickerman, Praxair, Inc., to S. S. Grossel (June 15, 1999).

Edwards, J. C. 1991. Thermal Models of a Flame Arrester. Bureau Mines Report RI 9378. U.S. Department of the Interior, Bureau of Mines, Washington, DC.

Enardo. no date. Brochure on Flame Arresters with Proven Performance. Enardo Manufacturing Company, Tulsa, OK.

FRS. 1985. Evaluation of the Performance of an End-of-Line Flame Arrester When Subjected to Unconfined Hydrogen–Air Explosions. Report of Investigation FRD 184/03 for IMI Amal Limited. Fire Research Station, Borehamwood, Herts., England.

FRS. 1990. Evaluation of the Performance of a Two Inch Group IIC Detonation Arrester When Tested to the Requirements of BS 7224: Report of Investigation FATR/1023 for IMI Amal Limited. Fire Research Station, Borehamwood, Herts., England.

Flessner, M. F. and Bjorklund, R. A. 1981. Control of Gas Detonations in Pipes. *CEP Loss Prevention Manual,* 14, 113–126. American Institute of Chemical Engineers, New York, NY.

Frobese, D. H. and Forster, H. 1992. Propagation of Detonations Through Pipework Junctions. *Proc. 7th Intl. Symposium on Loss Prevention and Safety Promotion in the Process Industries,* Taormina, Italy, May 4–8, 1992.

Gordon, S. and McBride, B. J. 1976. Computer Program for Calculation of Complex Chemical Equilibrium Compositions, Rocket Performance, Incident and Reflected Shocks, and Chapman-Jouguet Detonations. Report SP-273. Interim Revision, Report N78-17724. National Aeronautics and Space Administration, Washington, DC.

Green, D. W. and Maloney, J. O. 1997. *Perry's Chemical Engineers' Handbook.* 7th ed. pp. 6–12. McGraw-Hill, New York, NY

Grumer, J., Harris, M. E., and Rowe, V. R. 1956. *Fundamental Flashback, Blowoff, and Yellow-Tip Limits of Fuel Gas-Air Mixtures.* U.S. Bureau of Mines Report of Investigations 5225 (July 1956).

Halstrick, V. and Kirchner, W. 1996. Corrosion-Resistant Flame Arresters and Valves. Protego Technical Paper ha-sd 0660E780. DOC/8. Braunschweiger Flammenfilter GmbH, Braunschweig, Germany.

Howard, W. B. 1982. Flame Arresters and Flashback Preventers. *Plant/Operations Progress,* 1(4), 203–208.

Howard, W. B., Rodehorst, C. W., and Small, G. E. 1975. Flame Arresters for High-Hydrogen Fuel–Air Mixtures. *CEP Loss Prevention Manual,* 9, 46–53.

HSE. 1980. *Flame Arresters and Explosion Vents.* Health and Safety Series Booklet HS(G)11. Health and Safety Executive, London, England.

Kletz. T. A. 1982. A Flame Trap Assembly for Use with High Melting Point Materials. *Plant/Operations Progress,* 1(4), 252–254.

Kuchta, J. M., Cato, R. J., and Gilbert, W. H. 1970. Flame Arrester Materials for Fuel Tank Explosion Protection. Safety Research Center Report No. S4/38. Safety Research Center, Bureau of Mines, Pittsburgh, PA.

Langford, B., Palmer, K. N., and Tonkin, P. S. 1961. The Performance of Flame Arrestors Against Flame Propagating in Various Fuel/Air Mixtures. Fire Research Station Note No. 486. Fire Research Station, Borehamwood, Herts., England.

Lapp, K. 1982. Detonation Flame Arresters and Protected Side Explosion Venting. Paper presented at the 1992 ILTA Conference, June 23, 1992.

Lapp, K. and Vickers, K. 1992. Detonation Flame Arresters and Larger Line Sizes. International Data Exchange Symposium on Flame Arresters and Arrestment Technology. Banff, Alberta (October 1992).

Lapp, K. and Werneburg, H. 1999. How Flame Arresters Can Be a Hazard. Paper presented at the ILTA 19th International Operating Conference, June 14–17, 1999, Houston, TX.

Lapp, K. and Werneburg, H. 2000. Flame Propagation Research Identifies Vapour Handling Hazards. Paper presented at the ILTA 20th International Operating Conference, June 12–15, 2000, Houston, TX.

Lunn, G. A. 1982a. An Apparatus for the Measurement of Maximum Experimental Safe Gaps at Standard and Elevated Temperatures. *J. Hazardous Materials,* 6, 329–340 (1982).

Lunn, G. A. 1982b. The Influence of Chemical Structure and Combustion Reactions on the Maximum Experimental Safe Gap of Industrial Gases and Liquids. *J. Hazardous Materials,* 6, 341–359.

Magison, E. C. 1987. Hazardous Material Classification in the United States: History, Problems, and Needs. *Proc. Intl. Symposium on the Explosion Hazard Classification of Vapors, Gases and Dusts.* National Materials Advisory Board, Washington, DC.

Mendoza, V. A. 1999. Personal communication from Vicente A. Mendoza, NAO Inc., to S. S. Grossel (July 1999).

Mendoza, V. A., Smolensky, V. G., and Straitz, III, J. F. 1996. Don't Detonate—Arrest That Flame. *Chemical Engineering,* 103(5), 139–142.

NAS. 1975. Matrix of Electrical and Fire Hazard Properties and Classification of Chemicals. National Academy of Sciences Report to US Coast Guard, Contract No. DOT-CG-41680-A (1975).

NMAB. 1982. Classification of Gases, Liquids and Volatile Solids Relative to Explosion-Proof Electrical Equipment. Report NMAB 353-5. National Academy Press, Washington, DC (August 1982).

Overhoff, K-H, Schecker, H-G, Fellensiek, J., and Oncken, U. 1989. Investigation for the Design of a New Water Trap Flame Arrester. International Symposium on Loss Prevention and Safety Promotion in the Process Industries, Paper 54. Oslo, Norway, June 19–22, 1989.

Palmer, K. N. 1960. The Quenching of Flames by Perforated Sheeting and Block Flame Arresters. *Proc. Symposium on Chemical Process Hazards with Special Reference to Plant Design*, pp. 51–57. Institution of Chemical Engineers, Rugby, England.

Palmer, K. N. and Rogowski, Z. W. 1968. The Use of Flame Arresters for Protection of Enclosed Equipment in Propane–Air Atmospheres. *IChemE Symposium Series No. 25*, pp. 76–85. Institution of Chemical Engineers, Rugby, England.

Palmer, K. N. and Tonkin, P. S. 1963. The Quenching of Propane–Air Explosions by Crimped-Ribbon Flame Arresters. *IChemE Symposium Series No. 15*, pp. 15–20. Institution of Chemical Engineers, Rugby, England.

Phillips, H. 1981. The Differences between Determinations of Maximum Experimental Safe Gap. *J. Hazardous Materials*, 4(3), 245–256.

Phillips, H. 1987. A Comparison of "Standard" Methods for the Determination of Maximum Experimental Safe Gap (MESG). *Proc. International Symposium on the Explosion Hazard Classification of Vapors, Gases, and Dusts*, pp. 83–108. Publication NMAB-447. National Materials Advisory Board, Washington, DC.

Phillips. H. and Pritchard, D. K. 1986. Performance Requirements of Flame Arresters in Practical Applications. *IChemE Symposium Series No. 97*, pp. 47–61. Institution of Chemical Engineers, Rugby, England.

Piotrowski, T. C. 1991. Specification of Flame Arresting Devices for Manifolded Low Pressure Storage Tanks. *Plant/Operations Progress*, 10(2), 102–106.

Protego. 1993. *Special Catalogue of Protego Flame Arrester for Hydrogen Systems.* Protego Publication No. NO770993. Braunschweiger Flammenfilter GmbH, Braunschweig, Germany.

Rao, S. N., Dam, A. S. and Maus, F. G. 1980. Detonation Flame Arrester Testing for Oyster Creek Nuclear Station. Paper presented at the ANS/ENS Intl. Conference on World Nuclear Energy. Washington, DC, November 19, 1980.

Recklinghausen, K. 1978. Method and Devise for Protecting Ethylene Oxide Producing and Processing Plants Against the Decomposition of Ethylene Oxide. German Patent Submission No. P28.50254.7 (November 20, 1978).

Reed, R. D. 1977. Laminar Flow Flame Arrester. U.S. Patent No. 4,015,954 (April 5, 1977).

Reid, R. C., Prausnitz, J. M., and Sherwood, T. R. 1977. *The Properties of Gases and Liquids.* 3rd ed. pp. 544–601. McGraw-Hill, New York, NY.

Rogowski, Z. W. 1978. *Manual for Testing Flame Arresters.* Fire Research Station, Borehamwood, Herts., England.

Rogowski, Z. W. 1980. Flame Arresters in Industry. *IChemE Symposium Series No. 58,* pp. 53–65. Institution of Chemical Engineers, Rugby, England.

Rogowski, Z. W. and Pitt, A. I. 1976. Performance of a Two Element Crimped Ribbon Flame Arrester. Fire Research Station Note No. 1037. Fire Research Station, Borehamwood, Herts., England (January 1976).

Roussakis, N. and Brooker, D. E. 1995. U.S. Patent No. 5,415,233, *Arrester Apparatus* (May 16, 1995).

Roussakis, N. and Lapp, K. 1989. A Summary of Investigations from Ten Inline Flame Arrester Failures. Paper presented at a meeting of the API Committee on Safety and Fire Protection. Seattle, WA, September 21, 1989.

Saacke, F. C. 1963. Utilization of Arresters. *CEP Technical Manual on Acetylene Handling,* pp. 15–18. American Institute of Chemical Engineers, New York, NY.

Schampel, K. and Steen, H. 1975. Flame Arresting High Velocity Valves on Cargo Tanks of Tankers for Inflammable Liquids. *J. Hazardous Materials,* 1, 223–235.

Schmidt, H. 1971. Protective Measures and Experiences in Acetylene Decomposition in Piping and Equipment. *IChemE Symposium Series No. 34,* pp. 165ff. Institution of Chemical Engineers, Rugby, England.

Schneider, A. 2000. Personal communication from Dr. Alan Schneider, U.S. Coast Guard, to S. S. Grossel (April 12, 2000).

Schwartz, R. 1999. Personal communication from Robert Schwartz, John Zink Company, to S. S. Grossel (September 20, 1999).

Schwartz, R. 2000. Personal communication from Robert Schwartz, John Zink Company, to S. S. Grossel (March 13, 2000).

Smolensky, V. G. 1999. Personal communication from Vadim G. Smolensky, NAO Inc., to S. S. Grossel (August 4, 1999).

Straitz III, J. F. 1999. Personal communication from John F. Straitz III, NAO Inc, to S. S. Grossel (June 29, 1999).

Strehlow, R. A., Nicholls, J. A., Magison, E. C., and Schram, P. 1979. An Investigation of the Maximum Experimental Safe Gap Anomaly. *J. Hazardous Materials,* 3, 1–15.

Sutherland, M. E. and Wegert, H. W. 1973. An Acetylene Decomposition Incident. *Chemical Engineering Progress,* 69(4), 48–51.

Thibault, P., Britton, L. G., and Zhang, F. 2000. Deflagration and Detonation of Ethylene Oxide Vapor in Pipelines. *Process Safety Progress,* 19(3), 125–139 (Fall 2000).

Thomas, G. O. and Oakley, G. L. 1993. On Practical Difficulties Encountered When Testing Flame and Detonation Arresters to BS 7244. *Trans. Inst. Chem. Engr.,* 71, Part B, 187–193.

Tornado. no date. Brochure on Sure Stop Detonation Flame Arresters. Tornado Flare Systems, Stettler, Alberta, Canada.

Wauben, M. 1999. Personal communication from Matt Wauben, SchuF (USA), Inc., to S. S. Grossel (June 17, 1999).

Westech Industrial Ltd. no date. Flame Arrester Technical Presentation. Westech Industrial Ltd., Calgary, Alberta, Canada.

White, R. E. 2000. Personal communication from Robert E. White, Southwest Research Institute, San Antonio, TX, to S. S. Grossel (October 24, 2000).

Wilson, R. P. and Attalah, S. 1975. Design Criteria for Flame Contiol Devices for Cargo Venting Systems. U.S. Coast Guard Report CG-D-175-75. Department of Transportation, Washington, DC.

Wilson, R. P. and Flessner, M. F. 1978. Design Criteria for Flame Arresters. Paper presented at the 84th National AIChE Meeting, February 26–March 1, 1978, Atlanta, GA.

Zabetakis, M. G. 1965. *Flammable Characteristics of Combustible Gases and Vapors*. Bulletin 627. Bureau of Mines, U.S. Department of the Interior, Washington, DC.

Zanchetta, V. 1998. Personal communication from V. Zanchetta, The Kemp Company, to S. S. Grossel (October 10, 1998).

6

Installation in Process Systems

6.1. Design Considerations with Respect to Other System Components

When a flame arrester is installed in a system, consideration must be given to how it may adversely affect other system components. For example, when an end-of-line or in-line deflagration flame arrester is installed on the vent line of a low-pressure storage tank (designed per API 650 or UL Std. 142), it is very important to ensure that the flame arrester pressure drop is low enough so that the tank design pressure is not exceeded. Otherwise, the tank will be overpressured and may be deformed or possibly rupture, which could result in the release of hazardous chemicals. This may require the selection and installation of a larger size flame arrester, or it may also be necessary to oversize the vent nozzle (and any associated piping) to reduce the pressure drop.

Consideration should also be given to the possibility that the flame arrester may plug, which could produce a vacuum condition in a low-pressure tank when the tank is pumped out, and implode (collapse) the tank. This may require the installation of a vacuum breaker or a pressure-vacuum conservation valve. If the tank contents are flammable and admission of air may result in an ignitable mixture, it may be necessary to install an inert gas blanketing system on the tank, actuated by a pressure controller, which would admit a sufficient flow of inerting gas when a vacuum condition is detected.

For detonation flame arresters that are installed in manifolded vent systems, similar considerations as above apply to prevent overpressure or vacuum damage of atmospheric pressure tanks.

6.2. Piping and Flame Arrester System Design Considerations

When in-line detonation flame arresters are installed in single or manifolded vent systems, consideration must be given to the design of the system (flame arrester and piping). The following design practices should be considered:

1. Piping and flame arresters should be designed for the maximum anticipated temperature and pressure.
2. Piping supports should be designed to withstand the maximum anticipated pressure transients.
3. Proper materials of construction for the flame arrester and piping must be selected to minimize corrosion, which if uncontrolled, could possibly result in catastrophic failure and subsequent loss of containment.
4. Piping should be designed to avoid pocketing and liquid accumulation. If pocketing cannot be avoided, proper drainage should be provided at low points and the piping support system designed for the additional weight of accumulated liquid.
5. If there is a possibility of accumulated liquid freezing, the flame arrester and piping should be heat traced and insulated.
6. If a flame arrester is preceded by piping larger than the arrester inlet nozzle, it is necessary to install a pipe of at least 120 pipe diameters long of the same size as the flame arrester inlet nozzle upstream of the arrester (see Section 5.3.15).

Although the peak pressures associated with detonations are very high, they have very short duration (typically 1–2 ms). During this time the structural loading seen by the piping is minimal since the natural period of the piping components is not reached. However, it is important to ensure that the detonation flame arrester is designed and tested (hydrostatic or pneumatic pressure tests) to withstand the maximum line pressure that can occur. It is also critical to specify and install piping that will withstand the detonation. At initially low pressures of a few psig, and depending on factors which include size and configuration of the piping system, reactivity of the gas, strength and location of the ignition source, detonation velocities can approach 2000 m/s, and incident pressure (pressure on the side-wall of the pipe) can exceed 4000 psig. These incident pressures can be reflected from elbows or the ends of the piping system, or flat appurtenances perpendicular to the flow path of the gases and flame front in the piping, and the reflected shock can approach 9000 psig. Because the extremely high pressure portion of detonation loads are only for short periods of time, they only affect a short section of the piping at a given time. Because of the response of ductile materials to loads of short duration, designing a manifolded piping system for the peak pressures of a detonation is not required. Piping constructed of materials with a ductility of elongation ≥12% at rupture, a design tensile strength of ≥50,000 psig, and a design pressure rating of ≥350 psig should be able to contain detonations in systems which are operating at a maximum of 6 psig, and at temperatures between –20°F and 400°F (Henley 2000).

When a flame arrester is installed in a low pressure vapor collection manifold that may see high pressures during venting associated with an emergency shutdown condition (ESD), the flame arrester must be designed so that the housing is able to withstand this ESD venting situation. The same requirement applies to the manifold piping.

Fittings, elbows, and tees should be minimized since these affect flame propagation via the influence of flow turbulence and shock reflection. Experimental work by White and Oswald (1992) showed that elbows and fittings increase the flame arrester loading and may affect the arrester flame quenching capability. The capability of an end-of-line and in-line deflagration arrester will be exceeded, but a detonation arrester usually will be unaffected. Aside from the effect that the shock wave has on a flame arrester, the reaction of the pipe itself should be accounted for in the design of pipeline supports to withstand the bending-related forces and moments that are produced.

Piping manifolds should be designed to avoid pocketing and accumulation of transported and condensed liquids. They should be sloped so that the piping is inclined towards the low point. Some companies slope the piping toward a vessel in the system, or if this is not possible, then heat tracing is provided to prevent condensation from occurring at the low point. Low points in the piping should be provided with a means of drainage. It is very important to remove any accumulated liquid that can become a slug and impact on the flame arrester face if a deflagration or detonation occurs. Such impact can damage or plug the arresting element. Manual drainage can be considered if it is addressed by a formal procedure. Alternatively, it may be desirable to provide an automatic drain system at the low point to ensure the prompt draining of accumulated liquid.

For installations in cold climates, as well as applications in which product freezing, polymerization, or crystallization is a concern, the flame arrester and piping should be insulated and, and where required, heat traced. Where polymerization is a concern, the temperature of the tracing should be kept as low as possible. It is important to ensure that the temperature of the heat tracing be kept below the accepted operating range of the flame arrester. This can be achieved by using tempered water (often with glycol added), or low pressure steam. Electrical tracing with a temperature control system or self-limiting electric tracing is also often used.

6.3. Maintaining Reliability

Maintaining flame arrester reliability is critical to successful operation. Modifications to the system which may change gas composition should be carefully evaluated to ensure that the mixture with the lowest MESG is still within flame arrester certification. Other management of change issues include piping modifications, increased or decreased pressures, and the introduction of corrosive, condensable, or other potentially harmful materials.

To ensure reliability, dry-type flame arresters should be inspected on a regular basis, as mentioned previously. Also, as discussed in Section 5.3.18, it is often the practice to provide pressure drop and temperature instrumentation to monitor the flame arrester condition in service. This will provide indication of impending problems that can affect the performance of the flame arrester.

Hydraulic (liquid seal) flame arresters require attention on a regular basis similar to that for dry-type flame arresters. It is critical to ensure that the liquid level is at the required height, and level instrumentation with alarms is recommended. Automatic addition of makeup liquid (to replace evaporation and entrainment losses) is desirable. Temperature instrumentation is also recommended to monitor the occurrence of sustained burning (a stabilized flame). See Section 5.3.18 for more details.

6.4. Optimum Location in System

A number of factors should be considered when determining the optimum location of a flame arrester, such as:

- ease of inspection and maintenance
- minimization of run-up length to avoid high flame speed and failure of the flame arrester (flame breakthrough)
- emission of vent vapors which could impact on personnel

As mentioned in Section 5.3.16, one of the primary criteria for flame arrester location is that it be installed in a location that affords easy access for inspection and maintenance, including ease of removal. End-of-line flame arresters that are installed on outdoor tanks or process vessels should be mounted directly on the vent nozzle, and the vent nozzle should preferably be located near the edge of the tank roof for easy access. If the vent nozzle is not near the tank roof edge, the flame arrester should be an in-line deflagration arrester so that the emitted vapors do not impinge on the tank roof. For tanks or process vessels located inside buildings, the vent line should be routed above the roof, and an end-of-line flame arrester installed at the end of the vent pipe. In-line detonation flame arresters in manifolded vent systems may be mounted as needed for accessibility subject to piping constraints.

In-line deflagration flame arresters must have piping on the unprotected side as short as possible to minimize run-up length to avoid the potential for significant flame acceleration and possible DDT in the pipe. The run-up length is affected by pipe diameter, and the number and types of fittings, and not just piping straight length. The maximum allowable run-up length is strictly limited by the recommendations of the vendor. For Group D gases, the distance between the flame arrester and the end of the discharge pipe that vents directly into the atmosphere is sometimes no more than 20 feet. For gases in other NEC groups the allowable run-up length may be appreciably less, and it is strongly recommended that the flame arrester manufacturer be consulted for the maximum run-up lengths for their specific flame arresters. The allowed distance must be established by proper testing with the appropriate gas mixture and the actual pipe diameter to be used. Turbulence-promoting irregularities in the flow (bends, tees, elbows, valves, etc.) cannot be used unless testing has addressed the exact geometry.

The flame arrester discharge point should be located at a minimum height above grade, building roof, or an operating level, and pointed away from areas where personnel can be impacted by the vapors or flame. Generally, a height of seven to ten feet above any level where personnel can be exposed is sufficient. Nothing should be above the discharge point such as a platform, roof overhang, piperack, etc.

6.5. Supports for Static and Dynamic Forces

Vessel nozzles and piping connected to flame arresters must be designed to provide adequate support for static and dynamic forces. Static forces are primarily due to the weight of the arrester. Large end-of-line and in-line flame arresters can be quite heavy, so vessel nozzles must be strong enough to carry the weight of the arrester and to resist any bending moments. It may be necessary to reinforce the nozzle with gussets, or an independent support structure may be required. Some large in-line flame arresters may have to be installed horizontally, and the arrester and attached piping may need support to carry the weight and remove stresses on the piping.

Nichols (1999) describes a number of reaction forces that are generated during deflagrations and detonations in piping systems such as:

1. *A radial force on the pipe wall ahead of the deflagration wave.* There is a varying pressure between the acoustic wave and the flame front where the pressure builds from near atmospheric pressure, P_1 (step change at the wave front) to eight times P_1 (or higher) at the flame front. The pressure ratios depend on the flame acceleration. There is no such effect with a detonation.
2. *A local radial force on the pipe wall at the flame front where there is a step change in pressure and hence radial stress in the wall.* This pressure rises to about 8 times P_1 for a deflagration and about 20 times P_1 for a detonation. However, since the deflagration velocity is subsonic, the lower stress acts for a greater time compared to the higher stress for a detonation, and may or may not give greater potential pipe material damage.
3. *A radial force on the pipe wall behind the flame front.* For a closed pipe, this remains constant (at about $8P_1$) everywhere for a deflagration, but decays from about $20P_1$ to $8P_1$ behind the detonation over most of the pipe length.
4. *A pressure wave at bends which gives a longitudinal force acting to move the pipe assembly on the pipe hangers.* For a deflagration this reaction force is the pressure change times the pipe area. For a detonation there is an additional momentum effect which doubles the apparent pressure to about $40P_1$. The duration of the force depends on the pressure rise and decay profile, but is usually below the pipe assembly resonant frequency.
5. *Longitudinal out-of-balance forces.* For a Z-configuration pipe layout there is a steady force between the two bends generated by the pressure difference in the gas while the pressure front moves between the two bends. This gives a longer duration to the force which can approach the resonant frequency of the pipe.

Detonation flame arresters and associated piping must be able to withstand the effects of explosion transients that include the pressure or pressure-related force, the specific impulse (integral of the time versus overpressure) or net impulse (upstream minus downstream impulse) high-speed gas momentum transfer and flux, and temperature-, heat-, and thermal-flux related loads (White and Oswald 1992). White and Oswald conducted a series of experiments to obtain information on the effect of detonation flame arrester structural response to detonations traveling in a long pipe system, as would be encountered in marine vapor control systems. The work involved a theoretical engineering analysis, testing with a small-scale and large-scale pipe, and evaluation of the USCG test procedure. The small-scale tests were conducted in a 6-inch pipe using a USCG-approved detonation arrester and a simulated arrester (the arresting element was tightly wound wire cloth). The large-scale tests were conducted in an 18-inch pipe with a simulated detonation arrester (the arresting element was a section packed with 1-inch carbon steel Pall rings). The test gases were three concentrations of propane in air (4.3%, 6.5%, and 7.5%). Based on these tests, the following were concluded:

1. Overdriven detonations, not long-pipe stable detonations, provide a greater potential for mechanical damage to detonation flame arresters.
2. For describing structural loading functions needed for design analysis, the use of overdriven detonation data representing the net overpressure (run-up side pressure less protected side overpessure) on the flame arrester element and supporting structure is preferable to data representing only the run-up side, side-on overpressure. However, the run-up side transient history of side-on overpressure for overdriven detonations should provide a conservative estimate for design purposes.
3. For deflagrations, the dependence of overpressure and its effect on a flame arrester do not scale in proportion to pipe length, pipe diameter, or pipe length to diameter ratio.
4. There are no clearly discernable, broadly applicable, correlations between the 6-inch and 18-inch deflagration and detonation experiments. Therefore, comparisons were done on a parameter-by-parameter basis. However, comparisons of data taken during experiments with the two pipe sizes reveal that enough scale-related differences exist that interpolation between the two scales for an intermediate size should be done only where conditions are very similar. Then, overpressure and specific impulse can be estimated based on *L/D*.
5. Venting an explosion ahead of a flame arrester can reduce the thermal flux and the impulse to which the arrester is subjected. Test results indicate that peak side-on overpressure is halved, specific impulse is reduced by a factor of three, and the temperature is substantially reduced. However, overpressure and flame speed at the flame arrester do not appear to be changed significantly.

6. Based on testing and thermal modeling, the potential for reignition on the protected side of an approved flame arrester under detonation conditions is minimal. A detonation flame arrester that is subjected to a sustained, stable detonation through a very long pipe will receive a higher thermal flux than it would receive if subjected to an overdriven detonation. However, thermal modeling indicates that asymptotic conditions are obtained for some length of pipe which, if longer, will yield no significant increase in thermal flux.

White and Oswald present two computer programs in the report, which can be used to obtain estimates of the structural response and the thermal response of a detonation flame arrester. However, they point out that neither of the programs will reveal completely what is needed to design a successful flame arrester support. The authors also evaluated the USCG test procedure and present recommendations on how this test procedure should be modified to improve it.

6.6. References

6.6.1. Regulations, Codes of Practice, and Industry Standards

HSE (Health and Safety Executive). 1980. *Guide for Flame Arresters and Explosion Reliefs.* Booklet HS(G)11. Health and Safety Executive, Her Majesty's Stationery Office, London, England.

6.5.2. Specific References

Henley, R. 2000. Draft of proposed new NFPA Guide for Mitigation of Explosion Hazards in Manifold Piping Systems.

Howard, W. B. 1992. Use Precautions in Selection, Installation, and Operation of Flames Arresters. *Chemical Engineering Progress,* 88(4), 69–75.

Nichols, F. 1999. Personal communication from Fred Nichols (retired from ICI), Aston, Chester, England to S. S. Grossel (September 7, 1999).

White, R. E. and Oswald, C. J. 1992. Mitigation of Explosion Hazards of Marine Vapor Control Systems. Southwest Research Institute, Final Report, SWRI Project No. 06-4116, for the American Petroleum Institute, Washington, DC (October 1992).

7

Inspection and Maintenance of Flame Arresters

7.1. Need and Importance of Maintenance

Inadequate maintenance of flame arresters can result in their failure to perform as required. Failure can result from enlargement of the apertures due to corrosion, erosion, shock waves, or mishandling during cleaning. This results in the inability to stop the breakthrough of flames from deflagrations and detonations. Another cause for failure is pluggage due to entrainment of solids from the process, polymer deposits, freezing of water and organic liquids, and nests built by birds and insects. In this case, the plugged flame arrester may prevent or diminish vapor inflow and outflow from vessels, which can lead to vessel failure due to overpressure or vacuum collapse. Fixed element dry type flame arresters, such as end-of-line deflagration arresters, are especially susceptible to these failures. However, hydraulic (liquid seal) flame arresters, and other types can also fail to operate as required if regularly scheduled maintenance is not provided.

7.2. Mechanical Integrity Issues

This section discusses inspection and maintenance practices and procedures that affect the mechanical integrity of flame arrester installations.

7.2.1. Inspection

It is critical to the integrity of flame arresters that they be inspected and serviced on a regularly scheduled basis (preventive maintenance program) so that they continue to be effective.

The regularity of inspections and the time interval between inspections depends on the application and company practice. One flame arrester manufacturer recommends that inspection should be made a month after initial operation. Many companies perform inspection and maintenance at least once annually unless operating experience indicates that the inspection period should be changed. Some companies that continuously monitor the pressure drop do not inspect flame arresters until abnormal pressure drop is detected. However, pressure drop may not identify corrosion problems or absence of the element or gaskets. Even if pressure drop indicates no potential problem, it is recommended that all flame arresters be visually inspected on a scheduled basis. Plugging may not be readily detected by pressure drop where the normal vent flow is only a small fraction of the design basis flow.

Some flame arrester designs allow in-place inspection (via a removable housing cover plate); however, others have to be removed from the vent nozzle or in-line piping to inspect the flame arrester element.

There are some mechanical factors that sometimes hamper inspection. A flame arrester may be in a location that is not readily accessible or does not provide sufficient working space or mechanical aids. This is why, as was mentioned in Chapter 6, proper location and installation is very important. If removal of the flame arrester is required for inspection, then purging may be required before removal. This may be difficult to do easily because of the flame arrester construction or location. Also, some flame arresters may be constructed so that removal of the arrester element for inspection is not possible (i.e., the element is welded to the housing). Another problem is that the flame arrester is in a pipeline or directly attached to a vessel that is in constant use. In some cases, this might even be a problem during a period of a turnaround. In this situation it may be necessary to install two flame arresters in parallel with adequate valving to switch from one to the other and to provide for purging.

Before performing any inspection or maintenance on a flame arrester, the associated process equipment and piping should be taken out of service or isolated. The work area must be proven by test to be free of any harmful gases or vapors. It should also be verified that all piping is clean and free of obstructions and debris. All plant, company, local, state, and federal safety and fire codes and standards should be followed.

7.2.2. Current Maintenance Practices

Depending on the size of a flame arrester, the cleaning of the flame arrester element may be done in place or in the maintenance shop. Removal of the flame arrester from its location and transport to the maintenance shop may require heavy lifting equipment.

Care is needed to ensure that the cleaning method is compatible with the flame arrester in terms of the materials of construction and the arresters overall robustness. Suitable cleaning methods include the use of solvents, water, steam, compressed air, or ultrasonics.

For crimped metal ribbon elements, cleaning must be done carefully, because of the small and delicate flow channels involved. It is important not to clean the element with sharp objects that might open the flow channels or distort them, and thus disable the arrester. At the same time, the small flow channels may tenaciously hold solids filtered from the line. In this case it may be desirable to send the element to the flame arrester manufacturer for cleaning. It is advisable to stock spare elements for replacement of the element returned to the manufacturer for cleaning so that the flame arrester may be returned to service quickly.

If the flame arrester element is of the parallel plate type some manufacturers caution against disassembling the grid assembly to clean it. This is because such handling may increase the gaps between the plates to larger than the original design rendering the flame arrester ineffective.

It is critical upon reassembly to check that all gaskets are properly installed and that the element is correctly aligned in the housing to ensure that there is no bypassing of the element. Failure to do this may result in complete loss of protection.

On-line cleaning may be desirable for very large flame arresters as the elements may not be readily removed. One such on-line cleaning installation is provided for a 72-inch-diameter flame arrester assembly, which weighs more than 15,000 pounds and is installed in a 36-inch-diameter piping system (Anon 1999, Dafft 1999). The flame arrester is located in a vent gas collection system, upstream of an incinerator. The vent gas contains sticky solids that could eventually plug the flame arrester if not removed. To remove the sticky solids a cleaning assembly with multiple spray nozzles is provided for on-line cleaning of the flame arrester element with a mixture of steam and hot deionized water while the process gas continues to flow to an incinerator.

If a flame arrester element is damaged, its effectiveness will be impaired, and a replacement should be considered. It is important to ensure that the replacement meets all specifications and that it is installed correctly, per the manufacturer's recommendations. It is recommended that a flame arrester be inspected after it stops a flame (see Sections 5.3.16 and 5.3.17).

All flame arrester manufacturers have written instructions for installation, inspection, and maintenance, and these should be followed. Some flame arrester manufacturers require that maintenance be done at their facility to avoid invalidating the flame arrester warranty.

7.2.3. Documentation and Verification of Flame Arrester Maintenance

It is important that the inspection and maintenance activities be documented as required per company practice and any process safety management regulatory requirements. Maintenance records should indicate the inspection results and the scope of maintenance work performed, if any. Before the flame arrester is reinstalled in the process, it should be visually inspected to make sure that reassembly conforms to the manufacturer's recommendations. Some companies affix a tag to the flame arrester indicating the inspection and maintenance date.

7.3. Training and Competence Issues for Operating and Maintenance Personnel

Proper training on the operation and maintenance of flame arresters should be given to the appropriate personnel so that they will become competent with respect to the operating characteristics of the various types of flame arresters used in a plant, as well as their specific inspection and maintenance requirements.

Many flame arrester manufacturers provide training courses on inspection and maintenance of flame arresters, either at their facilities or a client's plant. It is recommended that such courses be scheduled for operating and maintenance personnel. Also, the OSHA PSM standard, 29 CFR 1910.119 (j)(3), should be consulted for requirements for training for process maintenance activities.

7.4. On-Stream Isolation and Switching of Parallel Spares

For vapor lines that cannot be shut down conveniently to remove and clean arrester elements, parallel flame arresters are sometimes installed. To be able to switch the flame arresters, suitable isolation valving must be installed upstream of the flame arresters, which will allow the switching to be done without interrupting operations. Care must be taken to assure that monitoring instrumentation and/or interlocks remain functional at all times. Procedures or interlocks must ensure that the process system is protected at all times.

7.5. Check List for Inspection

Flame Arrester Type	What to Look For
Dry fixed element	Plugging, corrosion, physical damage, correct bolt torque, contaminants, correct arrester orientation, functioning instrumentation
Hydraulic	Proper liquid level, seal fluid contaminants, foam, vessel corrosion, distributor pluggage, functioning instrumentation
Packed bed	Packing attrition, corrosion, fouling or plugging, distributor condition (for wet type), functioning instrumentation

7.6. References

7.6.1. Regulations, Codes of Practice, and Industry Standards

29 CFR 1910.119. *Process Safety Management of Highly Hazardous Chemicals.* U.S. Department of Labor, Washington, DC

7.6.2. Specific References

Anon. 1999. New Detonation Arrester Stops Flamefronts. *InTech*, p. 26 (July 1999).

Dafft, C. A. 1999. Personal communication from C. A. Dafft, Rohm and Haas Company, Deer Park, TX to S. S. Grossel (August 1999).

Howard, W. B. 1992. Use Precaution in Selection, Installation, and Operation of Flame Arresters. *Chemical Engineering Progress*, 88(4), 69–75.

8

Regulations, Codes, and Standards

This chapter discusses regulations, codes, and standards currently in use in the United States, Canada, the United Kingdom, and Europe for selecting, testing, and certification of flame arresters.

8.1. Regulations, Codes, and Standards Summaries

8.1.1. United States

In the United States flame arrester codes and standards have been published by the U.S. Coast Guard (USCG), the Underwriters Laboratories (UL), Factory Mutual Research (FM), and the American Society for Testing and Materials (ASTM). The USCG regulations are most widely used for testing and certification by flame arrester manufacturers. There are certain advantages to having UL listing or classification or FM approval (see Appendix C).

U.S. Coast Guard Regulations

The development of the U.S. Coast Guard (USCG) regulations (33 CFR Part 154) is discussed in Section 2.3.1. These regulations are mandatory only for maritime installations such as ship and barge loading and unloading facilities. However, some companies are applying some aspects of these regulations also for vent collection systems (manifolded systems).

In the development of these regulations, the USCG adopted findings from research on flame propagation and flame quenching in long pipes that was done at the Risk Analysis Laboratory in Calgary, Alberta, Canada, which was primarily funded by Westech Industrial Ltd. (a flame arrester manufacturer) and also some funding by Shell, Esso (Canada), and Chevron petroleum companies.

The USCG classifies deflagration flame arresters in two categories as follows:

Type I: Flame arresters acceptable for end-of-line applications. Where a Type I arrester is provided with cowls, weather hoods, or deflectors, etc., it shall be tested in each configuration.

Type II: Flame arresters acceptable for in-line applications. Type II arresters shall be specifically tested with the inclusion of all pipes, tees, bends, cowls, weather hoods, etc., which may be fitted between the arrester and the atmosphere. Owing to the prohibitive cost of testing deflagration flame arresters for each particular installation, the Type II (in-line) category is generally not encountered.

Appendix A to Part 154 of 33 CFR, Marine Vapor Control Systems, Final Rule (1990), contains Guidelines for Detonation Flame Arresters, including testing requirements. Detonation flame arresters are extensively tested for proof of performance against deflagrations, detonations, and endurance burns. Under this protocol, the test gas must be selected to have either the same or a lower MESG than the gas in question. Typical MESG benchmark gases are stoichiometric mixtures of propane, hexane, or gasoline in air to represent Group D gases having an MESG equal to or greater than 0.9 mm and ethylene in air to represent Group C gases with an MESG no less than 0.65 mm. Commercially available flame arresters are typically certified for use with one or another of these benchmark gas types. An ethylene-type flame arrester is selected should the gas in question have an MESG less than 0.9 mm but not less than 0.65 mm. Five low- and five high-overpressure deflagration tests are required with and without a flow restriction on the protected side. Of these 20 tests, the restricted-end condition is usually the more severe one and often limits the maximum initial pressure at which the flame arrester will be suitable.

The USCG test protocol for detonation flame arresters requires the following:

1. The flame arrester must be bidirectional.
2. The flame arrester tested must be able to withstand sequentially (a) five deflagrations with an outlet pipe length equal to 10 pipe diameters, (b) five deflagrations with a restriction in the outlet pipe

located 0.6 meters from the detonation flame arrester, (c) five stable detonations, and (d) five overdriven detonations. Section 14.3.4 of Appendix A outlines a trial-and-error method to determine the test conditions to generate the "maximum" overdriven detonation velocity that will be used in the test. It is important to note that the USCG protocol does not state minimum flame speeds and pressures for the deflagration testing. With regard to overdriven detonations, the USCG protocol attempts to ensure that detonation arresters are tested to the maximum severity (velocity and pressure) of overdriven detonations, but this cannot be guaranteed because it is extremely difficult to reproduce overdriven detonations in a testing environment which always behaves the same. This is pointed out by Lee et al. (1996), who propose a set of test configurations where all the required combustion regimes can be obtained reproducibly in a relatively short length of pipe, thus facilitating both the test apparatus and the test procedure.

3. If these tests are successful, an endurance burn test is required. If the detonation flame arrester can withstand 2 hours of continuous burning it is classified as a Type I flame arrester. If it can only withstand between 15 minutes to 2 hours of continuous burning it is classified as a Type II flame arrester. The USCG protocol does not clearly define the orientation of the detonation flame arrester (horizontal vs. vertical) during the continuous burn test. This makes a major difference. The burning test does not use propane for Group D gases, but hexane or gasoline, owing to their lower autoignition temperatures. For Group C gas tests, ethylene can be used for all test stages.
4. The detonation flame arrester must be able to withstand a hydrostatic pressure of 350 psig for 10 minutes without rupturing, leaking, or distorting.
5. After all flame tests the detonation flame arrester will be pneumatically tested to 10 psig to ensure that there is no leaking.

Appendix B to Part 154 is a standard specification for tank vent (end-of-line) deflagration flame arresters. It does contain test procedures for these types of flame arresters. An end-of-line flame arrester must be able to arrest three atmospheric deflagrations as well as to withstand two hours of continuous burning without flashback.

A USCG-accepted flame arrester must be of a configuration to prevent field modifications that might change its performance (e.g., the addition of a conservation vent, gooseneck, weatherhood, etc. which was not on the original flame arrester during testing).

Underwriters Laboratories Standard

Underwriters Laboratories Standard UL 525 (1994) applies to both deflagration and detonation flame arresters. It has also been approved as an ANSI standard. Its development is discussed in Section 2.3.1.

The UL 525 detonation flame arrester test criteria are as follows:

1. The detonation flame arrester can be either bidirectional or unidirectional. This is one of the main differences between the USCG and UL standards. However, as of April 2000 UL has only listed (approved) bidirectional flame arresters.
2. The detonation flame arrester must be able to arrest ten deflagrations with and without a pipe restriction downstream of the flame arrester and five unrestricted stable and overdriven detonations. The UL standard states, "after tests determine the maximum unstable (overdriven) detonation, the arrester is to be subjected to four additional unstable detonations with the length of pipe that resulted in the maximum unstable (overdriven) detonation. The arrester is also to be subjected to five stable detonations."
3. A UL Type I detonation flame arrester is one that will not flash back when subjected to an endurance burning. A UL Type II flame arrester is one that will flash back when subjected to an endurance burning test; it is marked with a continuous burn time (this is where UL 525 differs from the USCG test protocol). During the endurance burning test the concentration and the flow rate of the test mixture are varied while maintaining a flame on the surface of the arrester for at least 2 hours or until the temperature on the protected side has a temperature rise of at least 100°C (180°F). After 2 hours or a 100°C rise, the procedure is continued until the highest obtainable temperature is reached on the protected side of the arrester. The highest obtainable temperature is considered to have been reached when the temperature rise does not exceed 0.5°C (0.9°F) per minute over a 10-minute period, regardless of the concentration and flow rate. The highest obtainable temperature is maintained for 10 additional minutes without additional temperature rise after which the test mixture flow is stopped. In the continuous flame test the flow of the test mixture is stopped for 15 seconds every 10 minutes after initiation of the test. There shall be no flashback, as determined visually, at any time during the test, including the 15 seconds following the final 10-minute burning interval.
4. The detonation arrester must be hydrostatically tested to 350 psig for 10 minutes, without rupture, leakage, or permanent distortion.

The UL 525 test requirements for tank vent deflagration flame arresters are as follows:

1. The deflagration flame arrester must be subjected to a series of at least 10 explosion (deflagration) tests in a rig with a pipe at least 5 feet (1.5 meters) long with various mixtures of propane in air and different test conditions to test the entire spectrum of possible deflagrations. Also, a series of 3 flashback tests, using a mixture of 4.2 volume percent of propane in air, must be conducted.
2. A UL Type I deflagration flame arrester must undergo an endurance burn test while a UL Type II deflagration flame arrester must be subjected to a continuous flame test. The test conditions for the endurance burn test and the continuous flame test for a deflagration flame arrester are the same as for a detonation flame arrester.
3. The deflagration flame arrester must be hydrostatically tested to 250 psig for one minute without rupture or permanent damage.

Factory Mutual Research Standards

Factory Mutual Research (formerly Factory Mutual Research Corporation) has issued two flame arrester standards. One is a test procedure for flame arresters for storage tank vent pipes (end-of-line deflagration flame arresters). This test procedure (FMRC 1998) is primarily an endurance burning test in which the flame arrester is subjected to a natural gas-air mixture flame for a 30 minute burning period. After the initial 30 minute burning period, the room is darkened and the gas-air mixture is shut off for 10 seconds to see if flashback through the flame arrester occurs. If flashback does not occur, the gas-air mixture is then restarted and allowed to burn on the upper flame arrester element until the flame arrester element temperatures are approximately as they were at the end of the 30 minute burning period. The gas-air mixture is again shut off for 10 seconds. The procedure is conducted a total of three times on each flame arrester, and the flame arrester is accepted if there is no flashback.

The second standard is a procedure for examination and testing of detonation flame arresters (FMRC 1999). Testing to ascertain ability to stop deflagrations and detonations is conducted in accordance with the USCG requirements as outlined in 33 CFR, Appendix A to Part 154. However, the endurance burning test is conducted in accordance with the FMRC test procedure for flame arresters for storage tank vent pipes.

American Society for Testing and Materials (ASTM) Standard

ASTM published a standard specification F 1273 for tank vent flame arresters (end-of-line deflagration flame arresters) in 1991 (reapproved in

1996). This specification is intended for flame arresters protecting systems containing vapors of flammable or combustible liquids where vapor temperatures do not exceed 60°C (140°F).

Test procedures are as follows:

1. For vapors from flammable or combustible liquids with an MESG greater than or equal to 0.9 mm, technical grade hexane, technical grade propane, or gasoline vapors shall be used for all tests. For vapors with a MESG less than 0.9 mm, the specific vapor (or alternatively, a gas with a MESG less than or equal to the MESG of the specific vapor) shall be used for the tests.
2. Three tests must be performed and flame passage shall not occur during these tests in order for the flame arrester to be acceptable.
3. An endurance burning test is required where the flame arrester is heated until the highest obtainable temperature on the upstream side of the flame arrester is reached. The highest obtainable temperature may be considered to have been reached when the rate of temperature increase does not exceed 0.5°C per minute over a 10 minute period. This temperature shall be maintained for a period of ten minutes, after which flow shall be stopped and the conditions observed. Flame passage shall not occur during the endurance burning test for the arrester to be acceptable.

Flame arresters meeting this specification also comply with the minimum requirements of the International Maritime Organization (IMO), Maritime Safety Committee Circular No. 373 (MSC/Circ. 373/Rev. 1). This specification is not widely used in the US chemical industry (Pietrowski 2000).

8.1.2. Canada

The Canadian Standards Association Standard Z343 (CSA 1998) presents test methods for in-line and firebox flame arresters. In this standard in-line flame arresters are limited to only detonation types and firebox flame arresters are defined as flame arresters installed in an enclosure, or system of enclosures, where the run-up distance is less than 1.5 meters and open to the atmosphere. Firebox flame arresters are commonly used on equipment designed to heat fluids in production operations such as indirect heaters, emulsion treaters, and glycol dehydrators. The development history of this standard is presented in Section 2.3.2.

The standard presents procedures for conducting burn tests for in-line flame arresters, and deflagration and detonation test methods for in-line flame arresters and firebox flame arresters. In the burn tests a propane–air mixture is used for Class I, Group D applications, and an ethylene–air mixture is used for Class I, Group C applications. The test flame arrester is subjected to the flame until 3 hours have elapsed, or until flame propagation occurs to the protected side of the element. One burn test each shall be conducted at velocities of 15, 30, 90, and 180 meters/minute, and the time to flashback shall be recorded for each test velocity. For the in-line detonation flame arresters, the following tests are required: 5 restricted-end stable detonations, 5 restricted-end overdriven detonations, 15 restricted-end deflagrations (at low, medium, and high pressure). For firebox flame arresters, the tests are conducted with four ignition points located at 50 mm or less, 1, 2, and 2.5 meters from the flame arrester connection to the test pipe. Each firebox flame arrester shall be subjected to at least five ignitions at each ignition location. There shall be no flame propagation through the flame arrester in any of the deflagration and detonation tests for the flame arrester to be acceptable.

8.1.3. United Kingdom

The British Standards Institute standard specification BS 7244 (1990) applies to both deflagration (end-of-line and in-line) and detonation flame arresters. For end-of-line deflagration flame arresters ten tests are required, and for in-line deflagration flame arresters fifteen tests are required. For detonation flame arresters three tests at increasing lengths of pipe for both deflagration and detonation conditions and ten unrestricted overdriven detonation tests are required. Endurance burning test procedures are presented, but tests are conducted only if specified.

There are some practical difficulties with testing of deflagration and detonation flame arresters in accordance with BS 7224 as pointed out by Thomas and Oakley (1993), based on their experience while establishing a facility to test flame arresters in accordance with this standard. The procedures given in this standard were found to be insufficient to develop a system whose test parameters could be easily compared with other facilities. Great difficulty was found in generating transition to detonations in propane–air mixtures in a smooth-walled tube. Transitions to detonation were observed in ethylene–air mixtures, but the process was highly stochastic in nature and was only observed when an energetic oxyacetylene booster section was used as an initiator. Based on their experiences they suggested several things to consider when the standard was to be revised. However, when the new CEN flame arrester standard (see Section 8.1.4 below) is issued, it will supersede BS 7224.

8.1.4. Europe and International

CEN European Standard

A new CEN European standard, EN 12874, was issued in 2001. It will be used by all European nations that belong to the European Union (EU) that are members of the Comité Européean de Normalisation (CEN), as well as other non-EU nations that are members of the CEN. It will supersede BS 7224 in England, PTB and BAM in Germany, and standards in use in other European countries. This is a very comprehensive standard covering types of flame arresters not covered in US, Canadian, and other standards. It contains test procedures for the following types of flame arresters:

1. static flame arresters (fixed element dry type)
2. liquid product detonation flame arresters (in liquid filled or partially filled liquid piping)
3. high velocity vent valves
4. flow controlled apertures (velocity flame stoppers)
5. hydraulic flame arresters (liquid seals for gas/vapor lines)
6. flame arresters combined with breather valves (conservation vent valves)

It also contains specific requirements for testing of flame arresters in equipment such as compressors, blowers, fans, and vacuum pumps. Procedures are presented for deflagration, detonation, and burning tests for all the flame arrester types in the standard.

The CEN standard has some significant differences and/or clarifications in comparison to the USCG, FM, and UL standards, such as:

1. A detonation flame arrester can be tested and approved for either stable detonations only or for both stable and unstable (overdriven) detonations. Three tests for stable detonations and five tests for unstable detonations must be conducted. In addition, three deflagration tests must be conducted. In each case various lengths of piping are installed downstream of the flame arrester and are blocked off with a blind flange. For detonation testing, the downstream piping shall have a length of 10 times the diameter and not less than three meters. For the deflagration tests the downstream piping shall be 50 times the diameter for hydrocarbon-air mixtures and 30 times the diameter for hydrogen-air mixtures. Approved detonation arresters are also considered approved for arresting deflagrations. Hydrostatic testing is required for prototypes.
2. In-line flame arresters for both detonations and deflagrations and end-of-line flame arresters can be tested and approved for atmo-

spheric deflagrations and endurance burning, or for atmospheric deflagration and short-time burning. Endurance burning must be done in the orientation to be used in service. Otherwise the testing is similar to US standards.

3. With regard to in-line flame arresters the CEN standard clearly specifies minimum operating pressures as well as detailing of the operating pressure on the test report.
4. The CEN procedure for in-line deflagration flame arresters requires careful recording of pipe run-up distances to help reduce confusion about how far from the flame source the deflagration arrester can be installed. In these tests the pipe downstream of the flame arrester is equal in length to the pipe upstream of the flame arrester and is blocked off on both the upstream and downstream side with a blind flange. Six deflagration tests are required. Hydrostatic testing is not required.
5. There are test procedures for liquid product flame arresters, high velocity vent valves, flow controlled apertures (velocity flame stoppers), and hydraulic flame arresters, as well as specific requirements for testing flame arresters used with compressors, fans, blowers, and vacuum pumps, which are not covered in other standards.
6. The CEN standard dictates the use of temperature sensors on flame arresters not approved for continuous (endurance) burning.
7. The CEN standard uses the European classification system for flammability which has more groups (based on the MESG of the vapor–air mixture) than the US classification system.
8. The CEN standard specifically addresses the issue of endurance burn tests.

According to Wauben (1999) the CEN standard offers many differences and advantages to flame arrester users, such as:

1. It clearly defines the capabilities and testing conditions of a flame arrester instead of giving an all-encompassing approval. This is especially true for in-line deflagration flame arresters.
2. It allows more flexibility to engineers with regard to the type of flame arrester used and its approval, thus helping to possibly minimize problems such as maintenance, plugging, and pressure drop.
3. Test procedures for liquid product detonation arresters, hydraulic flame arresters, high velocity vent valves, and flow controlled apertures (velocity flame stoppers) allow the use of these kinds of flame arresters as an alternative to standard static fixed element dry type flame arresters. This could be especially useful in high flow applica-

tions, dirty applications where flame arrester plugging could be a major problem, or in corrosive environments where a plastic-lined hydraulic flame arrester could be more economical.

4. The more comprehensive CEN standard makes it likely that an approved device is available for almost any application. In certain difficult applications it may be required to make modest design changes in order to accommodate the flame arrester and its capabilities, but this is preferable to being stuck with an application where no approved flame arresters are available.

Thomas (1998) presents an overview of the contents of the CEN standard and comparison of it with other existing test standards.

Germany

In Germany flame arresters are tested in accordance with the TRbF standard (Technical Rules for Inflammable Liquids) and approved by the German Federal Physical-Technical Institute (PTB) and the German Federal Institute for Material Research and Testing (BAM). However, this standard will be superseded by the new CEN standard.

International Maritime Organization (IMO) Standard

The IMO standard MSC/Circ. 677 (1994) provides testing procedures for end-of-line deflagration and in-line detonation flame arresters, as well as high velocity vent valves for use on cargo tanks in tanker ships. Its development is discussed in Section 2.3.4.

The test gas may be propane, hexane, or gasoline vapors. For end-of-line deflagration flame arresters, tests shall be performed twice each for three ignition sources, for a total of six tests. For in-line detonation flame arresters three detonation tests are required.

8.2. Comparison of Various Flame Arrester Standards and Codes

Table 8-1 presents a comparison of test requirements in various US and foreign standards and codes for end-of-line and in-line deflagration flame arresters. A comparison of test requirements in various US and foreign standards and codes for in-line detonation flame arresters is similarly presented in Table 8-2. The UL and FM test criteria closely follow the USCG criteria.

TABLE 8-1

Comparison of Test Parameters of Various Standards for End-of-Line and In-Line Deflagration Flame Arresters (Adapted from Lapp and Werneburg 1999)

Standard	Test Rig	Test Gas	Deflagrations	Test Pressure	Endurance Burn Test
		END-OF-LINE DEFLAGRATION FLAME ARRESTERS			
USCG	Tank and plastic bag	Propane or specific gas	3	Atmospheric	Yes
BSI	Tank and plastic bag	Propane, Ethylene, hydrogen	10	Atmospheric	Yes, if specified
UL 525	Pipe, at least 1.5 m long	Propane	10 at various gas concentrations	As specified	Yes, plus continuous burn
IMO	Tank and plastic bag	Propane, Hexane, gasoline	6	Atmospheric	Yes
CSA			NO TEST		
FM	1.2 m vertical pipe	Propane or specific gas	0	Atmospheric	Yes, if required
CEN	Tank	Propane, ethylene, hydrogen	6	Atmospheric	
PTB		NO INFO AVAILABLE			
		IN-LINE DEFLAGRATION FLAME ARRESTERS			
USCG	Tank, including all piping and fittings	Propane or specific gas	3	Atmospheric	Yes
BSI	Straight pipe for flame quench	Propane, ethylene, hydrogen	15	Atmospheric	Yes, if specified
UL 525			NO TEST		
IMO			NO TEST		
CSA			NO TEST		
FM			NO TEST		
CEN	Straight pipe, closed, <50 *L/D*, at least 3m	Propane, Ethylene, hydrogen	6	As specified	Yes, if required
PTB			NO INFO AVAILABLE		

TABLE 8-2
Comparison of Test Parameters of Various Standards for In-Line Detonation Flame Arresters (Adapted from Lapp and Werneburg 1999)

Standard	Test Pipe	Test Gas	Stable Detona-tion	Over-driven Detona-tion	Deflagra-tions	Test Pressure	Endur-ance Burn Test
USCG	A	Propane or specific gas	5 unrestr.	5 unrestr.	10 unrestr. 10 restr.	As specified	Yes
BSI	A	Propane Ethylene Hydrogen	3 unrestr.	11 unrestr.	3 unrestr.	Atmos-pheric	Yes, if req.
UL 525	A	Propane	5 unrestr.	5 unrestr.	10 unrestr. 10 restr.	As specified	Yes
IMO	A	Propane	3 unrestr.	0	0	Atmos-pheric	No
CSA	A	Propane Ethylene	5 restr.	5 restr.	15 restr.	As specified	Yes
FM	A	Propane or specific gas	5 unrestr.	5 unrestr.	10 restr. 10 unrestr.	As specified	Yes
CEN	B	Propane Ethylene Hydrogen	3 closed end	5 closed end	3 closed end	As specified for stable and unstable detona-tions	Yes, if req.
PTB	B	Propane Ethylene Hydrogen	3 closed end	0	0	As specified	No

A denotes straight pipe for stable detonations
B denotes straight pipe, closed end for stable detonations
restr. denotes a pipe with a restricted end
unrestr. denotes a pipe with an unrestricted end

8.3. Standards and Codes in Preparation

At the present time there are no plans for revising the UL, FM, and ASTM standards. However, the USCG is presently revising its regulations relative to flame arresters.

8.4. References

8.4.1. Regulations, Codes of Practice, and Industry Standards

ASTM F-1273-91. 1991. *Standard Specification for tank Vent Flame Arresters.* American Society for Testing and Materials, West Conshohocken, PA.

BSI (British Standards Institution) BS 7244. 1990. *Flame Arresters for General Use.* British Standards Institution, London, England, UK.

CEN (Comité Européean de Normalisation) EN 12874. 2001. *Flame Arresters—Specifications, Operational Requirements and Test Procedures.* European Committee for Standardization, Brussels, Belgium.

CSA (Canadian Standards Association) CSA-Z343. 1998. *Test Methods for In-Line and Firebox Flame Arresters.* Canadian Standards Association, Toronto, Ontario, Canada.

FMRC (Factory Mutual Research Corporation) Class 6061. 1990. *FMRC Test Procedure for Flame Arresters for Storage Tank Vent Pipes.* Factory Mutual Research Corporation, Norwood, MA.

FMRC (Factory Mutual Research Corporation) Class 6061. 1999. *Examination Program, Detonation Flame Arresters.* Factory Mutual Research Corporation, Norwood, MA.

IMO (International Maritime Organization) MSC/Circ. 677. 1994. *Revised Standards for the Design, Testing, and Locating of Devices to Prevent the Passage of Flame into Cargo Tanks in Tankers.* International Maritime Organization, London, England, UK.

33 CFR Part 154, Subpart E-Vapor Control Systems. 1990. U.S. Department of Transportation, Washington, DC.

UL 525. 1994. *Standards for Flame Arresters.* 6th edition. Underwriters Laboratories, Inc., Northbrook, IL.

8.4.2. Specific References

Lapp, K. and Werneburg, H. 1999. How Flame Arresters Can Be a Hazard. Paper presented at the ILTA 19th Annual International Operating Conference, June 14–17, 1999, Houston, TX.

Lee, J. H. S, Knystautas, R., and Goroshin, S. 1996. The Testing of Detonation Arresters. *Proc. Intl. Symp. On Hazards Prevention and Mitigation of Industrial Explosions*, pp. 7.27–7.40. Christian Michelsen Research AS, Bergen, Norway.

Piotrowski, T. 2000. Personal communication from Tom Piotrowski, Protectoseal Company, Bensenville, IL (January 4, 2000).

Thomas, G. O. and Oakley, G. L. 1993. On Practical Difficulties Encountered When Testing Flame and Detonation Arresters to BS 7244. *Trans. IChemE,* 71, Part B, 187–193.

Thomas, G. O. 1998. Explosion Arrester Testing: State of the Draft European Standard. Paper presented at the 3rd Worldwide Seminar on Explosion Phenomena and Application of Explosion Protection Techniques in Practice, February 8–12, 1998, Europex, Ghent, Belgium.

Wauben, M. 1999. Personal communication from Matt Wauben, SchuF (USA), Inc., Mt. Pleasant, SC, to S. S. Grossel (June 17, 1999).

9

Illustrative Examples, Calculations, and Guidelines for DDA Selection

9.1. Introduction

This chapter presents a number of examples and calculations that illustrate the selection, application, and sizing of DDAs and other protective measures (discussed in Chapter 3) for the prevention of flame propagation. Also included is a list showing the factors that should be considered for the selection of an appropriate DDA.

9.2. Example 1—Protective Measures for a Vent Manifold System

This example illustrates several protective measures that were provided for a vent manifold system in an actual aromatics chemical plant.

Figure 9-1 is a schematic drawing of the major equipment and protective measures that comprise the vent manifold system. In the system shown, the vent vapors (offgas) from the condensers of two vacuum columns are collected in a manifold which goes to the vacuum pump system. From the vacuum pump system, the vapors go to a seal drum (hydraulic flame arrester), and then to the firebox of a process heater, where they are incinerated.

The offgases from the two vacuum columns, at the vacuum pump discharge, normally contain 1.3 mole percent organic vapor (C_6 to C_8 aromatics) in air, which slightly exceeds the lower flammable limit of the

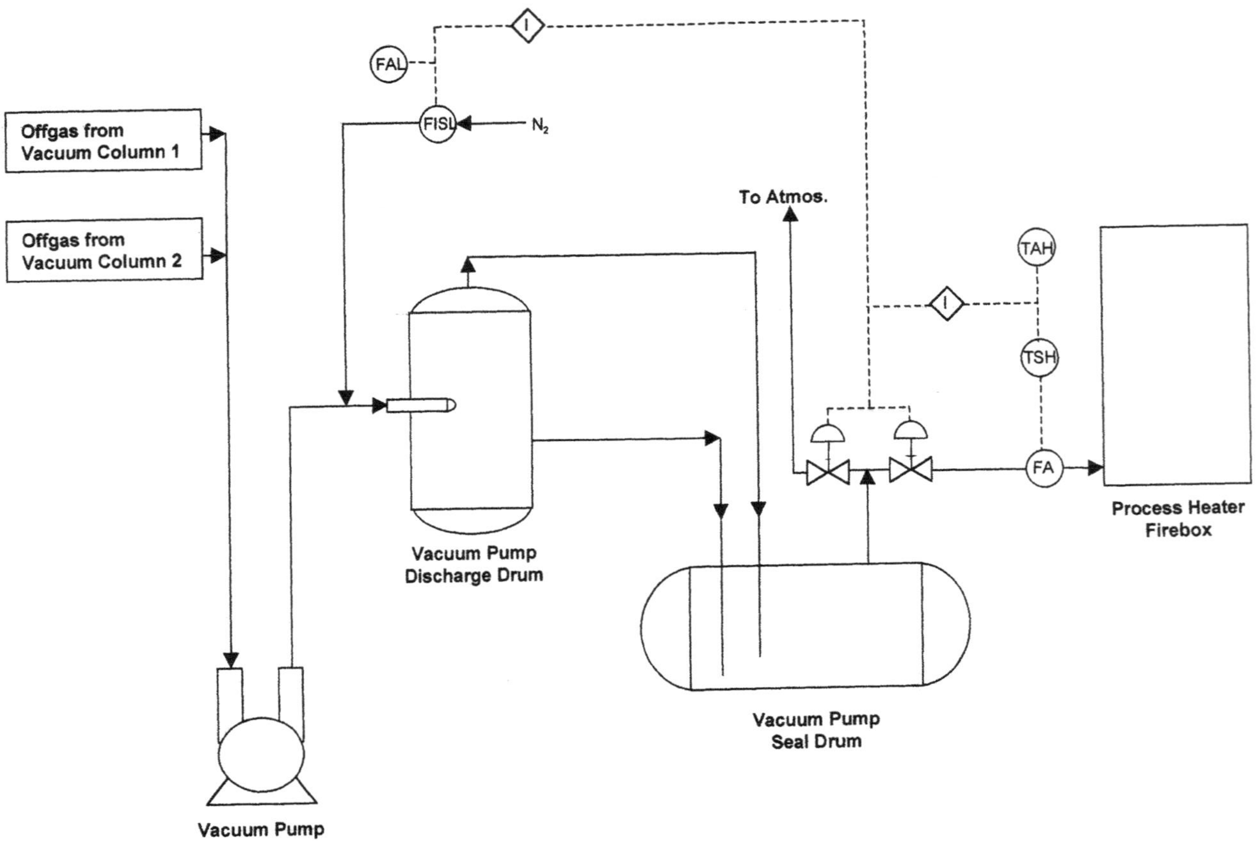

FIGURE 9-1. Vent manifold system protective measures.

mixture (about 1.0 mole percent organic in air). This stream, therefore, requires constant inerting with nitrogen for safe injection into the process heater firebox flame zone.

The following safety measures were provided:

- A constant nitrogen addition into the discharge line from the vacuum pump to the vacuum pump discharge drum/seal drum system.
- A vacuum pump seal drum design which provides a liquid seal (hydraulic flame arrester) to mitigate flame propagation backward into the vacuum system. The seal liquid is an organic stream (mostly C_8 aromatics) that comes from the vacuum pump discharge drum overflow.
- A detonation flame arrester with an integral thermocouple at the inlet to the process heater firebox to prevent backflash into the vacuum system.
- An interlock system (sensors and valves) which isolates offgas flow to the process heater firebox and routes the offgas to atmosphere on detection of low nitrogen flow or high temperature at the detonation flame arrester outlet.

This example illustrates the use of several protective measures to minimize the possibility of flame propagation.

9.3. Example 2—Sizing of an End-of-Line Deflagration Flame Arrester

This example shows the calculations necessary to size an end-of-line deflagration flame arrester.

Size an end-of-line deflagration flame arrester for the normal vent nozzle of a 126,000 gallon API-type atmospheric pressure storage tank for the following conditions:

Liquid stored: methyl alcohol (flash point = 52°F)
Liquid pump-in rate: 250 gpm
Liquid pump-out rate: 200 gpm
Tank design pressure: 6 inches WC positive and 2 inches WC negative
Liquid storage temperature: ambient

Venting capacity for liquid movement and thermal breathing are based on API Std. 2000 (1998).

Calculations

1. **Calculate the outbreathing rate (for a combination of liquid pump-in and thermal breathing).**
   Provide 1 SCFH of air per every 3.5 gph of maximum filling rate for liquids with flash points of less than 100°F.

$$\text{SCFH air} = \text{gpm} \times 60/3.5 + T_v$$

where T_v is the thermal venting capacity (see Table 9-1)

$$\text{SCFH air} = 250 \times 60/3.5 + 3000 = 4285.7 + 3000 = 7285.7$$

TABLE 9-1
Required Venting Capacity in SCFH (Interpolation Permitted)

TANK CAPACITY		VACUUM	PRESSURE	
GALLONS	42 GAL. BARRELS	ALL STOCKS	FLASH PT. BELOW 100°F.	FLASH PT. ABOVE 100°F.
2,500	60	75	75	—
5,000	119	125	125	60
10,000	238	250	250	137
15,000	357	350	350	200
20,000	476	480	480	270
25,000	595	600	600	350
30,000	714	730	730	425
35,000	833	850	850	500
42,000	1,000	1,000	1,000	600
84,000	2,000	2,000	2,000	1,200
126,000	3,000	3,000	3,000	1,800
168,000	4,000	4,000	4,000	2,400
210,000	5,000	5,000	5,000	3,000
420,000	10,000	10,000	10,000	6,000
630,000	15,000	15,000	15,000	9,000
840,000	20,000	20,000	20,000	12,000
1,050,000	25,000	24,000	24,000	15,000
1,260,000	30,000	28,000	28,000	17,000
1,470,000	35,000	31,000	31,000	19,000
1,680,000	40,000	34,000	34,000	21,000
1,890,000	45,000	37,000	37,000	23,000
2,100,000	50,000	40,000	40,000	24,000
2,520,000	60,000	44,000	44,000	27,000
2,940,000	70,000	48,000	48,000	29,000
3,360,000	80,000	52,000	52,000	31,000
3,780,000	90,000	56,000	56,000	34,000
4,200,000	100,000	60,000	60,000	36,000
5,049,000	120,000	68,000	68,000	41,000
5,880,000	140,000	75,000	75,000	45,000
6,720,000	160,000	82,000	82,000	50,000
7,560,000	180,000	90,000	90,000	54,000

Source: Protectoseal Bulletin.

2. Calculate the inbreathing rate (for a combination of liquid pump-out and thermal breathing).

Provide 1 SCFH of air per every 7.5 gph of maximum emptying rate for liquids of all flashpoints.

$$\text{SCFH air} = \text{gpm} \times 60/7.5 + T_v$$

$$\text{SCFH air} = 200 \times 60/7.5 + 3000 = 1600.0 + 3000 = 4600.0$$

3. Select the flame arrester size from a vendor's flow curve for the largest flow rate.

Figure 9-2 is a flow curve (tank pressure vs. air flow), from one flame arrester manufacturer, that indicates pressure drop for flow in either direction. This figure illustrates that a 3-inch flame arrester would be adequate.

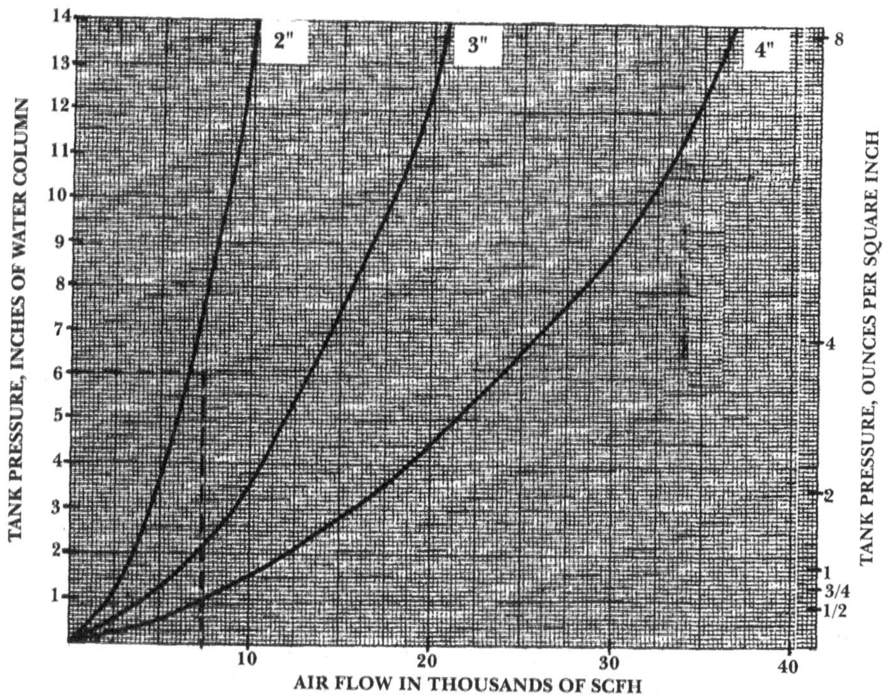

FIGURE 9-2. Typical flow curves for end-of-line deflagration flame arresters.

9.4. Example 3—Calculation of Limiting Oxidant Concentration (LOC)

This example shows how to calculate the limiting oxidant concentration of a vapor if an experimentally determined value is not available.

From Eq. (4-2) (page 57), the LOC of a vapor, when nitrogen is the diluent, can be calculated as follows:

$$\text{LOC} = \text{LFL(moles oxygen/moles fuel)}$$

Calculation

Calculate the LOC of n-butane, as follows:

1. The LFL of n-butane is 1.6 volume percent in air (Crowl and Louvar 1990).
2. The amount of oxygen needed to completely react with butane is:

$$C_4H_{10} + 6.5\ O_2 \rightarrow 4\ CO_2 + 5\ H_2O$$

3. LOC = 1.6(6.5 moles O_2/mole n-butane) = 10.4 volume percent oxygen.
4. The actual LOC, as reported in NFPA 69, is 12 volume percent oxygen when using nitrogen as an inerting gas. This calculation, in general, gives a value that is on the conservative (safer) side.

9.5 Example 4—Calculation of the LFL and UFL of Mixtures

This example shows how to calculate the LFL and UFL of gas–vapor mixtures, using the Le Chatelier rule (Crowl and Louvar 1990).

The LFL and UFL of a mixture can be calculated by the following equations:

$$\text{LFL}_{\text{mix}} = \frac{1}{\Sigma(y_i/\text{LFL}_i)}$$

$$\text{UFL}_{\text{mix}} = \frac{1}{\Sigma(y_i/\text{UFL}_i)}$$

where LFL_i is the lower flammable limit for component *i* in volume percent of component *i* in fuel and air; UFL_i is the upper flammable limit for component *i* in volume percent of component *i* in fuel and air; and y_i is the mole fraction of component *i* on a combustibles only basis.

Calculation

Calculate the LFL and UFL of a gas mixture composed of 0.8% hexane, 2.0% methane, and 0.5% ethylene by volume in air.

The data for this calculation are listed below.

Component	Volume percent	Mole fraction on combustible basis	LFL (vol. %)	UFL (vol. %)
Hexane	0.8	0.24	1.1	7.5
Methane	2.0	0.61	5.0	15.0
Ethylene	0.5	0.15	2.7	36.0
Total combustibles	3.3			
Air	96.7			

$$LFL_{mix} = \frac{1}{0.24/1.1 + 0.61/5.0 + 0.15/2.7} = 1/0.396 = 2.53\%$$

by volume total combustibles

$$UFL_{mix} = \frac{1}{0.24/7.5 + 0.61/15.0 + 0.15/36.0} = 1/0.0769 = 13.0\%$$

by volume total combustibles

Since this mixture contains 3.3% by volumne total combustibles, it is above the LFL_{mix}, and below the UFL_{mix}, thus is within the flammable range.

9.6. Example 5—Calculation of the MESG of Mixtures

This example illustrates how to calculate the MESG for a gas–vapor mixture using the original Le Chatelier rule, which includes only the MESG values of the combustible components, and the modified Le Chatelier rule (proposed in NFPA 497), which includes inert gases as well.

The equation for the MESG of a gas/vapor mixture is as follows:

$$MESG_{mix} = \frac{1}{\Sigma(y_i / MSEG_{mix})}$$

where y_i is the mole fraction of component i in the mixture, and $MESG_i$ is the MESG of component i

Calculation

Calculate the MESG of the following mixture using both the original and the modified Le Chatelier rule:

Component	Volume Percent	MESG, mm
Ethylene	45	0.65
Propane	12	0.97
Nitrogen	20	∞
Methane	3	1.12
Isopropyl ether	17.5	0.94
Diethyl ether	2.5	0.83

Using the original Le Chatelier rule the mixture MESG will be calculated using only the combustible gases (i.e., ignoring the nitrogen).

The composition of the gas mixture then is as follows:

Component	Volume Percent
Ethylene	56.25
Propane	15.00
Methane	3.75
Isopropyl ether	21.88
Diethyl ether	3.12

The mixture MESG is then equal to:

$$\left(\frac{0.5625}{0.65}+\frac{0.15}{0.97}+\frac{0.0375}{1.12}+\frac{0.2188}{0.94}+\frac{0.0312}{0.83}\right)^{-1}=\frac{1}{1.3239}=0.755\text{ mm}$$

Since this MESG value is close to the upper value for Group C (greater than 0.45 mm and less than or equal to 0.75 mm), it might be prudent to classify the mixture as Group C, or consider testing the mixture for a more accurate determination.

Using the modified Le Chatelier rule, the mixture MESG is equal to:

$$\left(\frac{0.45}{0.65}+\frac{0.12}{0.97}+\frac{0.20}{\infty}+\frac{0.03}{1.12}+\frac{0.175}{0.94}+\frac{0.025}{0.83}\right)^{-1}=0.944\text{ mm}$$

Since the estimated mixture MESG is greater than 0.75 mm, the mixture is considered as a Group D gas.

It can thus be seen that using the original Le Chatelier rule may result in specifying a flame arrester with smaller apertures. Using the original Le Chatelier rule for flame arresters may be the more the appropriate thing to do (see Section 5.3.1).

9.7. Determination If a DDT Can Occur

It is not always easy to determine when a DDT may occur in a piping system, and when to specify either a deflagration or detonation flame arrester. Chatrathi et al. (2001) state that, based on their experimental work, when the *L/D* ratio in a straight pipe reaches a value of 60 to 70, a DDT will occur.

9.8. Typical Locations in Process Systems

It is very important to install deflagration and detonation flame arresters in the correct location. This is discussed in detail in Chapter 6, but some considerations and recommendations are presented below.

End-of-line Deflagration Flame Arresters

Figure 3-1 (page 19) shows where an end-of-line deflagration flame arrester should be located. It preferably should be installed on a nozzle that is near the outer edge of the tank/vessel roof for ease of accessibility for maintenance and removal.

In-line Deflagration Flame Arresters

Figure 3-2 (page 20) shows the installation of an in-line deflagration flame arrester. It also preferably should be located on a nozzle that is near the outer edge of the tank/vessel roof. The length of the vent discharge piping should be kept as short as possible to avoid the possibility of a DDT occurring. Recommendations for the maximum length of the vent discharge piping should be obtained from flame arrester manufacturers. This length may vary depending on the design of the flame arrester and the vapor or gas being handled (e.g., Group B, C, or D).

In-line Detonation Flame Arresters

Figure 3-3 (page 21) shows where detonation flame arresters should be located in a vent manifold (vapor collection) system and upstream of a vapor control unit. The detonation flame arresters should be installed on nozzles near the outer edge of the tank/vessel roof, similarly to the deflagration flame arresters discussed above. If a detonation flame arrester is located upstream of an item of equipment with an open flame in it, such as a thermal oxidizer, it should be located at a distance far enough from the inlet nozzle to avoid impingement on the arrester element from the open flame or radiant heat. Flame arrester manufacturers should be consulted for their recommendations on how far away from the flame source (e.g., thermal oxidizer) to locate the detonation flame arrester.

9.9. List of Steps in the Selection of a DDA or Other Flame Propagation Control Method

In the selection of an appropriate DDA or other flame propagation control method consideration must be given to a number of factors which affect the choice. Presented below is a list of items to be considered in a logical sequence.

1. Is a flammable mixture possible in a vessel and/or piping system?
2. If it is, does ignition in one part of the system need to be contained/isolated from other parts of the system?
3. If it does, what is the most appropriate way of doing this, i.e., deflagration venting, oxidant concentration reduction (inerting or gas enrichment), combustible concentration reduction (dilution), deflagration suppression, deflagration pressure containment, equipment and piping isolation, or installation of a DDA?
4. If a DDA is chosen, the following factors should be considered and data obtained:

(a) normal and upset conditions of flow, pressure, temperature, and composition
(b) type of DDA (dry type, hydraulic, packed bed, etc.)
(c) gas/vapor MESG and/or NEC Group
(d) physical location of the DDA and whether it should be a deflagration of detonation flame arrester
(e) allowable pressure drop (using a 20% fouling allowance)
(f) should the DDA be unidirectional or bidirectional?
(g) can a stabilized flame (continuous burn) be established on the DDA?
(h) materials of construction/corrosion effects of the gas or vapor
(i) is a thermocouple integral with the DDA required?
(j) are pressure drop taps required?

(k) are nozzles needed for injection of cleaning media?
(l) is piping upstream of the DDA larger than the DDA?
(m) is certification (USCG, UL, FM, CEN, etc.) required?
(n) what other vendor options are available or required?

9.10. References

9.10.1. Regulations, Codes of Practice, and Industry Standards

API Std. 2000. 1998. *Venting Atmospheric and Low-Pressure Storage* Tanks: Nonrefrigerated and Refrigerated. 5th Edition. American Petroleum Institute, Washington, DC.

NFPA 497. 1997. *Recommended Practice for the Classification of Flammable Liquids, Gases, or Vapors and of Hazardous (Classified) Locations for Electrical Installations in Chemical Process Areas.* National Fire Protection Association, Quincy, MA.

9.10.2 Specific References

Chatrathi, K., Going, J. E., and Grandestaff, B. 2001. Flame Propagation in Industrial Scale Piping. Proc. AIChE 35th Annual Loss Prevention Symposium, April 22–26, 2001, Houston, TX.

Crowl, D. A. and Louvar, J. F. 1990. *Chemical Process Safety: Fundamentals with Applications.* Prentice Hall, Englewood Cliffs, NJ.

10

Summary

Research conducted since the 1970s in the United States, Canada, the United Kingdom, and Germany has significantly advanced knowledge of the factors affecting the design and performance of deflagration and detonation flame arresters. However, there are still many areas where further information is required, and experimental work is needed. This chapter summarizes the state-of-the-art at publication time of this book and recommends other information and research work needed to fill in the gaps for the areas where knowledge is lacking.

10.1. Status of DDA Technology

It is not possible to design DDAs from first principles alone. Current state-of-the art requires that all potential DDA selections be based on the process fluid (gas or vapor) involved, piping and equipment layout and design, and potential process operating conditions. Based on the research that has been done to date and testing protocols that are accepted and being used, the following listed information and criteria summarize the status of current DDA technology (CCPS 1993, Lapp 2000, Lapp and Werneburg 2000, White and Oswald 1992):

- Flame arresters are categorized by the MESG of the gas being handled and its corresponding classification by the National Electrical Code (NEC) groups in the US or the International Electrotechnical Commission (IEC) groups in Europe (see Chapter 5).

- If a flame arrester is not certified as meeting the above conditions, it may not function as required (see Chapter 5).
- For mixtures with component MESGs greater than the test gas, using a flame arrester that has been certified for any gas or vapor with a lower MESG is conservative (safe) (see Chapter 5).
- A flame arrester must not be installed in pipes larger than those that the arrester has been tested for. When a flame may arrive from a pipe with a diameter larger than the arrester connection, the arrester should be installed no closer than 120 pipe diameters from the larger pipe. This allows transient large overpressures and turbulence produced by the reduction in pipe diameter to dissipate (see Chapter 5).
- A protected-side restriction, such as a valve, can severely reduce a flame arrester's capabilities by magnifying reflected initial pressure on the arrester element. In such a situation, a downstream restriction should be included in the test protocol (see Chapter 6).
- Many test protocols are not application-specific and do not provide the user with a clear idea of how to select a flame arrester (see Chapter 8).
- Overdriven detonations, not long-pipe stable detonations, provide a greater potential for mechanical damage to detonation flame arresters (see Chapter 6).
- Fittings, such as elbows and tees, will affect flame propagation in various ways. The number of elbows in a piping system will affect the performance of a deflagration flame arrester, but will not affect the performance of a detonation flame arrester. Two important effects are manifested by flow turbulence and shock reflection. Testing shows that fittings will increase pressure and velocity impacts on a flame arrester. Pipe fittings and valves create turbulence causing the confined flame front to accelerate in a shorter distance than in a straight run of pipe. In this short distance a flame front can reach speeds and pressures beyond the capability of end-of-line and in-line deflagration flame arresters, and they will fail. Contrary to expectations, elbows located close to a deflagration flame arrester are significantly more hazardous (higher likelihood of flame passage) than elbows located closer to the ignition point (see Chapter 5).
- For deflagrations, the effect of overpressure on a flame arrester does not scale in proportion to pipe length, pipe diameter, or pipe length to diameter ratio (see Chapter 6).

10.2. Recommended Practices

- A flame arresting device accepted by standards setting organizations should only be used within the range of testing parameters used for acceptance, including operating pressure, vapor composition, and pipe length. Deviations from test parameters may result in flame penetration through the arrester. In the United States, the acceptable testing protocols are the USCG, FM, and UL test standards. In Canada, CSA-Z343 is followed, and in Europe, CEN Standard EN 12874 will replace all existing other European national standards (see Chapter 8).
- If a flame arrester is provided with cowls, weather hoods, deflectors, etc., it must be tested for the configuration involved if the test is done to meet the UL, USCG, or CEN standard (see Chapter 8).
- Maintenance of a flame arrester should be performed carefully to avoid any adverse impact on arrester performance (see Chapter 7).
- All flame arresters should be inspected regularly as operating experience dictates (see Chapter 7).
- Continuous monitoring of pressure drop and temperature should be carried out. Pressure drop should be monitored if fouling and subsequent plugging is suspected or has previously occurred. Temperature monitoring should be provided if it is possible for a standing flame to occur on the flame arrester face and subsequently destroy the element (see Chapter 7).
- After a flame arrester has experienced an abnormal process event, internal inspection should be made and the element replaced if it has been damaged (see Chapter 7).
- For describing structural loading functions needed for design analysis, the use of overdriven detonation data representing the net overpressure (run-up side less protected side overpressure) on the arrester element and supporting structure is preferable to data representing only the run-up side, side-on overpressure. However, the run-up side transient history of side-on overpressure for overdriven detonations should provide a conservative estimate for design purposes (see Chapter 6).
- The reaction of the pipe itself should be accounted for in the design of the pipeline supports because of bending-related forces and moments that are produced (see Chapter 6).
- The length of installed pipe between the ignition source and the flame arrester is critical to the proper performance of end-of-line and in-line deflagration flame arresters. Line lengths greater than those used in the test allow higher flame speeds and pressure to

develop, causing the arrester to fail. For detonation flame arresters, lengthening the pipe between the ignition source and the flame arrester will not create any higher peak flame speed or pressures. The flame quenching ability of the arrester will not be affected (see Chapter 5).

- An approved flame arrester should not be misapplied, i.e., an end-of-line arrester should not be installed in an in-line application.

10.3. Why Flame Arresters Fail

A deflagration or detonation flame arrester fails by definition if any flame propagates from the unprotected to the protected side. Failures can result for a number of reasons, some of which are listed below:

- misapplication (including "grandfathered" arresters)
- incorrect design (wrong MESG, NEC group, flow rate data, etc.)
- excessive operating pressure (greater than test pressure)
- longer pipe runs than used in test setup
- more fittings than used in test setup
- mechanical failure (element and/or housing)
- fouling (dirty gas/vapor, polymerizable vapor, etc.)
- freezing (gas/vapor contains water or other material that can solidify)
- larger pipe size upstream of arrester inlet nozzle
- corrosion of element and/or housing
- unexpected endurance burning (not instrumented or tested for)
- protected-side restrictions that were not included in test
- deflagration arrester installed in pre-volume vessel system
- effect of heat tracing and insulation on quenching capacity (reduces heat sink)
- bypassing of arrester via bypass piping or instrumentation (e.g., differential pressure impulse lines)
- inadequate maintenance

10.4. Future Technology Development

Although an appreciable amount of information is currently available about the factors affecting the design, performance, and installation of deflagration and detonation flame arresters, there is still a lack of information on many other relevant factors. A number of suggested test procedures and research projects that would help to fill in the missing data are presented below (CCPS 1993, Britton 2000, Patenaude 2000, White and Oswald 1992):

1. There is a need to have all accepted test standards require that deflagration testing be done with a restricted outlet. More research is needed to evaluate the effect of restriction size and location on flame arrester effectiveness. Not all test standards require this for deflagration testing. This is important since protected side restrictions severely reduce flame arrester capability.
2. Tests should be performed to determine the effect of varying the gas composition on standard restricted-end deflagrations (RED) results. The present protocols do not require this, nor do they require varying the ignition location despite evidence that this affects flame arrester performance.
3. There is a need to develop a standardized test approach to induce an overdriven detonation since the overdriven detonation has the potential of producing the most severe mechanical damage to a flame arrester.
4. There is a need to standardize endurance burning protocols.
5. Work should be done to:
   (a) determine the effect of MESG and AIT on DDA effectiveness,
   (b) evaluate methods for determining MESGs for mixtures from pure component data,
   (c) evaluate methods for determining the effect of nonflammable (inert) components on mixture MESGs,
   (d) develop a test apparatus for determining MESG data that correlate DDA performance more reliably than afforded by existing apparatuses.
6. Tests are needed to determine the effects of multicomponent lean and rich mixtures on the performance of deflagration and detonation flame arresters. Combustion of lean mixtures can result in spin and galloping detonations which have more focused and higher pressures, and thus are of greater concern with respect to the structural integrity of flame arresters and other pipeline devices (e.g., fast-closing valves). Lean mixtures are more prevalent than stoichiometric mixtures in most manifolded vent systems.
7. Tests are needed to determine the effects of temperature (above and below ambient) on the performance of deflagration and detonation flame arresters.
8. Tests are needed to verify the design criteria of nonstandard flame arresters (hydraulic flame arresters, packed bed arresters, etc.).

10.5. References

Britton, L. G. 2000. Using Maximum Experimental Safe Gap to Select Flame Arresters. *Process Safety Progress*, 19(3), 140–145 (Fall 2000).

CCPS. 1993. *Guidelines for Engineering Design for Process Safety*. Center for Chemical Process Safety, American Institute of Chemical Engineers, New York, NY.

Lapp, K. 2000, Personal communication from K. Lapp, Westech Industrial Ltd., to S. S. Grossel (July 12, 2000).

Lapp, K. and Werneburg, H. 2000. Flame Propagation Research Identifies Vapour Handling Hazards. Paper presented at the ILTA 20th International Operating Conference, June 12–15, 2000, Houston, TX.

Patenaude, E. 2000. Personal communication from E. Patenaude, Westech Industrial Ltd., to S. S. Grossel (June 28, 2000).

White, R. E. and Oswald, C. J. 1992. Mitigation of Explosion Hazards of Marine vapor Control Systems. Southwest Research Institute, Final Report, SWRI Project No. 06-4116, for the American Petroleum Institute, Washington, DC (October 1992).

APPENDIX

A

Flame Arrester Specification Sheet for Manufacturer Quotation

To enable flame arrester manufacturers to select and provide the correct flame arrester for a specific application, they must be given all relevant data pertaining to the proposed installation. These data are usually transmitted to manufacturers by means of a flame arrester specification (data) sheet. When in-line detonation flame arresters are to be installed in a vented manifold (vapor collection) system, it is imperative that a system sketch (or piping isometric drawing) be included with the specification sheet to show all horizontal and vertical piping runs (with dimensions) and all fittings (elbows, tees, valves, etc.). These are needed to estimate the potential for DDT and to determine where all detonation arresters must be located in the piping. Also, if the gas/vapor stream is a multicomponent mixture, a complete definition of the stream composition, including minor corrosive or fouling components, should be appended to the data sheet. The MESG, NEC group, and AIT for each component should be listed as well.

Printed on the next page is a flame arrester specification (data) sheet that is a composite of data sheets from several flame arrester manufacturers and CPI companies. Most flame arrester manufacturers have their own data sheets, which can be used with additional data added as the user deems necessary.

FLAME ARRESTER SPECIFICATION (DATA) SHEET

Customer____________________________ Date____________

Project No._______________ Flowsheet No.__________________

Flame Arrester No.__________ Plant Site______________________

Process (Application) Data

Gas/Vapor______________________ Composition__________________

Normal Flow Rate__________________ Max. Flow Rate_________________

Normal Operating Press._____________ Max. Operating Press.____________

Normal Operating Temp._____________ Max. Operating Temp.____________

Molecular Weight__________________ AIT_________________________

NEC Group______ IEC Group________ MESG________mm

Max. Allowable Pressure Drop________________

Is Continuous Burning Possible on FA Element ☐ Yes ☐ No

Distance To Be Installed from Potential Ignition Source________

Applicable Test Protocol: ☐ UL 525 ☐ USCG ☐ FM ☐ CEN ☐ Other

Listing/Approval: ☐ UL Listed or Classified ☐ FM Approved

Additional Information____

Design Data

Type: ☐ Detonation FA
☐ In-Line Deflagration FA ☐ End-of-Line Deflagration FA
☐ Hydraulic (Liquid Seal)
☐ Removable Element ☐ Nonremovable Element
☐ Unidirectional ☐ Bidirectional

Installation: ☐ Horizontal ☐ Vertical ☐ Other

Pipe Size/Schedule_________ Flange Press. Rating/Face__________

Materials of Construction:
Housing____________ Element____________ Flange_____________

Options:
Pressure Tap______________________________
No./Size/Type

Temperature Probe Fitting___________________________
No./Size/Type

Drain Plug_____________________________
No./Size/Type

Coating/Special Paint_________________________________

Other Options______________________________________

Manufacturer Deliverables

☐ Certified Drawings ☐ Certified Flow Curves ☐ Weight
☐ Calculations ☐ Installation and Maintenance Instructions

APPENDIX

B

List of Flame Arrester Manufacturers

Listed below are the names, addresses, telephone numbers, fax numbers, and websites of flame arrester and deflagration suppression system manufacturers whose units are available in the United States as of the date of the publication of this book. This list is not intended to be all-inclusive and is not intended to be an endorsement of any manufacturer or its products.

Enardo Manufacturing Company
4470 South 70th East Avenue
Tulsa, OK 74145-4607
Tel: (918) 622-6161; (800) 336-2736 Fax: (918) 622-0004
Website: www.enardo.com

ESAB Welding & Cutting Products
L-Tec Steel Industry Products Division
P. O. Box 100545
411 Ebenezer Road
Florence, SC 29501-0545
Tel: (843) 664-2160 Fax: (843) 664-4462
Website: www.esab.com

Fenwal Safety Systems
700 Nickerson Road
Marlborough, MA 01752
Tel: (508) 481-5800 Fax: (508) 485-3115
Website: www.fenwalsafety.com

Fike Protection Systems
704 S. Tenth Street
Blue Springs, MO 64013
Tel: (816) 229-3405 Fax: (816) 229-4615
Website: www.fike.com

Groth Corporation
1202 Hahlo Street
P. O. Box 15293
Houston, TX 77220-5293
Tel: (713) 675-6151 Fax: (713) 675-6739
Website: www.grothcorp.com

IMI AMAL Limited
Holdford Road
Witton
Birmingham B6 7ES, England, United Kingdom
Tel: (44) 121-356-2000 Fax: (44) 121-356-7987
Website: www.imiamal.co.uk
U.S. Affiliate
IMI Safety Systems
1325 South Creek Drive, Suite 300
Houston, TX 77084
Tel: (281) 492-8377 Fax: (281) 492-8801
Website: www.imiamal.com

KEMP
4647 S.W. 40th Avenue
Ocala, FL 34474
Tel: (352) 873-5123 Fax: (352) 873-5124
Website: www.cmkemp.com

KnitMesh Limited
Sanderstead Station Approach
South Croydon
Surrey CR2 OYY, England, United Kingdom
Tel: (44) 181-651-6321 Fax: (44) 181-651-4095
Website: www.knitmesh.co.uk

Matheson Gas Products
166 Keystone Drive
Montgomeryville, PA 18936
Tel: (215) 641-2700 Fax: (215) 641-2714
Website: www.mathesongas.com

NAO Inc.
1284 Sedgley Avenue
Philadelphia, PA 19134
Tel: (215) 743-5300; (800) 523-3495
Fax: (215) 743-3018; (215) 743-3020
Website: www.nao.com

Oseco Inc.
1701 W. Tacoma
Broken Arrow, OK 74012
Tel: (918) 258-5626 Fax: (918) 251-2809
Website: www.oseco.com

Protectoseal Company
225 W. Foster Avenue
Bensenville, IL 60106
Tel: (630) 595-0800; (800)323-2268 Fax: (630) 595-8059
Website: www.protectoseal.com

Protego
Braunschweiger Flammenfilter GmbH
Industriestrasse 11
D-38110 Braunschweig, Germany
Tel: (49) 53 07 809-090 Fax: (49) 53 07 78 24
Website: www.protego.de
U.S. Representative
SchuF (USA), Inc.
486 Long Point Road
Mt. Pleasant, SC 29464
Tel: (843) 971-1162 Fax: (843) 881-6121
E-mail: SchuF_USA@msn.com

Rembe GmbH Messe-und Regeltechnik
P. O. Box 1540
D-59918 Brilon, Germany
Tel: (49) 2961-7405-10 Fax: (49) 2961-507 14
Website: www.rembe.de
U.S. Representative
Cv Technology, Inc.
11320 Fortune Circle, Suite G18
West Palm Beach, FL 33414-8742
Tel: (561) 753-9556 Fax: (561) 753-9563
Website: www.cvtechnology.com

Rexarc
P. O. Box 7
West Alexandria, OH 45381
Tel: (937) 839-4604 Fax: (937) 839-5897
Website: www.rexarc.com

Selas Corporation of America
P. O. Box 200
Dresher, PA 19025
Tel: (215) 283-8313; (800) 523-6500 Fax: (215) 646-3536
Website: www.selas.com

Shand & Jurs
L&J Technologies Company
5911 Butterfield Road
Hillside, IL 60162
Tel: (708) 236-6000 Fax: (708) 236-6006
Website: www.ljtech.com

Tornado Flare Systems (Canada) Inc.
Box 1390
Stettler, Alberta T0C 2L0, Canada
Tel: (403) 883-2400; (800) 661-4128 Fax: (403) 883-2550
Website: www.tornadotechnologies.com

Varec Vapor Recovery
10800 Valley View Street
Cypress, CA 90630-5016
Tel: (714) 761-1300 Fax: (714) 952-2701
Website: www.varec.com

Westech Industrial Ltd.
5636 Burbank Crescent S.E.
Calgary, Alberta T2H 1Z6, Canada
Tel: (403) 252-8803 Fax: (403) 253-6803
Website: www.westech/ind.com

Western Enterprises
875 Bassett Road
Westlake, OH 44145
Tel: (440) 871-2160; (800) 783-7890 Fax: (440) 835-8283
Website: www.westernenterprises.com

John Zink Company
11920 E. Apache
Box 21220
Tulsa, OK 74121-1220
Tel: (918) 234-1800; (800) 421-9242 Fax: (918) 234-2700
Website: www.johnzink.com

APPENDIX

C

UL and FM Listings and Approvals

Presented below is information on how UL and FM listings and approvals are obtained.

Underwriters Laboratories Inc. (UL) and Factory Mutual Research (FM) will conduct tests for flame arrester manufacturers to ascertain if a designated flame arrester (deflagration or detonation type) will prevent passage of a flame of a specific gas. Both UL and FM test procedures also include tests for endurance burning, and in the case of UL tests, also for continuous flames.

UL will usually run the tests at their own test facilities in accordance with UL 525. When a flame arrester successfully passes the tests it is given a "Listing" mark and is so published in their Gas and Oil Equipment catalogue. UL will also witness tests at other test facilities, if a flame arrester manufacturer so wishes, under the UL "Witnessed Test Data Program." If the tests are conducted in accordance with UL 525, then the flame arrester is listed similarly as if the tests had been done at UL's laboratory. If a witnessed test is done at another laboratory in accordance with another test protocol (not UL 525), then the flame arrester is designated as "Classified." All flame arresters that have a UL Listing or Classified mark may be found on their web site: www.ul.com

FM will conduct tests at their own test facilities usually only for deflagration flame arresters, following their Class 6061 test procedure. They will also observe and approve tests done at a flame arrester manufacturer's own test facility for both deflagration and detonation type flame arresters. For detonation flame arresters FM will accept tests that are done in accordance with the USCG test procedure. When a designated flame arrester has passed the prescribed tests, it is listed as "Approved" in the Factory Mutual Research Approval Guide. Lists of FM-approved flame arresters can be obtained from FM upon request.

DDAs which are UL listed/classified or FM approved are contained in a published list. In addition to passing the basic approved/testing protocols, FM approved or UL listed/classified devices will have follow-up inspections/reviews to ascertain that the tested design has not been modified, and that the devices are constructed according to the original design. USCG approved devices are not listed and have no follow-up service.

APPENDIX

D

Suggested Additional Reading

Listed below are suggested additional reading (articles and reports) that are not cited in the main text of this book.

Abrahamsen, A. R. no date. The Use of Flame Arresters in Incineration and Flare Systems. IMI Amal technical paper. IMI Amal Limited, Birmingham, England

Bishop, K. and Knittel, T. 1993. Do You Have the 'Right' Flame Arrester in Service? *Hydrocarbon Process., Intl. Ed.*, 72(2), 63–66.

Britton, L. G. 1997. Subsection on Flame Arresters, pp. 26-38 to 26-43, Section 26—Process Safety in *Perry's Chemical Engineers' Handbook,* edited by Green, D. W. and Maloney, J. O., 7th ed., McGraw-Hill, New York, NY.

Capp, B. 1991. Flame Arresters and Their Use in Bulk Transfer. *Proc. Conf. on Fire and Explosion Hazards,* Section 7, pp.183–190.

Capp, B. 1994. Flame Arresters: Endurance Burning. *IChemE Sympos. Ser.* No. 134, pp. 405–414.

Chatrathi, K. 1992. Deflagration Protection of Pipes. *Plant/Operation Progress,* 11(2), 116–120.

Cooper, S. P., Moore, P. E., Capp, B. and Senecal, J. 1992. Investigation into the Use of Active Detonation Arresters for Solvent and Waste Gas Recovery Systems. *IChemE Sympos. Ser.* No. 130, pp. 385–409.

Dainty, E. E. and Lobay, G. 1991. An Investigation of Flame Arrester Continuous Burn Test Protocols. Tech. Rept. MRL-91-136, Canada Center for Mineral and Energy Technology (CANMET), Ottawa, Ontario (December 1991).

Dodge, F. T., White, R. E., and Bankler, T. 1998. Risk Assessment of Liquid Seals in Marine Vapor Control Systems. Final Report, SWRI Project No. 04-8954, Southwest Research Institute, San Antonio, TX (June 1998).

Edwards, K. L. and Norris, M. J. 1999. Materials and Constructions Used in Devices to Prevent the Spread of Flames in Pipelines and Vessels. *Materials and Design,* 20(5), 245–252.

Forster, H. 1993. Comparative Analysis of Test Protocols for Flame Arresters. *Proc. Marichem '93 Conf.,* Session 8, Paper 1, November 30–December 2, 1993, Amsterdam, The Netherlands.

Forster, H. 2001. Flame Arresters—The New Standard and the Consequences. Paper presented at the Intl. European Safety Management Group Sympos.-Process Safety and Industrial Explosion Protection, March 27–29, 2001, Nurnberg, Germany.

Groman, J. 2000. Applying Detonation Flame Arresters. Paper presented at the 20th Annual ILTA Conference, June 12–15, 2000, Houston, TX.

HSE. 1996. *Flame Arresters.* HSE Book HS(G)158. Health and Safety Executive, available from HSE Books, P. O. Box 1999, Sudbury, Suffolk CO10 6FS, England.

Knittel, T. 1992. Flame Arresters for Industrial Use. Technical Paper 040, Westech Industrial Limited, Calgary, Alberta, Canada.

Knittel, T. 1993. Battery Site Vapor Control for Sour Oil Production Sites, Paper presented to various petroleum production companies. Westech Industrial Limited, Calgary, Alberta, Canada.

Khitrin, L. N. et al. 1965. Peculiarities of Laminar- and Turbulent-Flame Flashbacks. *Proc. 10th Sympos. (Intl.) on Combustion,* pp. 1285–1291.

Lapp. K. 1991. Marine Vapor Control, USCG Standards, Momentum, MESG, and What Next? Paper presented at the ILTA Sustaining Member Trade Show, June 19–20, 1991, Houston, TX.

Lapp, K. 1992. Passive Detonation Flame Arresters. Paper presented at the API 1992 Marine Technical-Environmental Conference, January 8–9, 1992, Chantilly, VA.

Lapp, K., Thibault, P., Ward, S., and Weiler, D. 1991. The Effect of Momentum and Thermal Flux in Long Lines on the Westech Second Generation Detonation/Flame Arrester. Paper presented to DOT/USCG Hazardous Material Branch Bulk Cargo Section, June 12, 1991, Washington, DC.

Lapp, K. and Vickers, K. 1992. The Effect of Reductions in Vapour Collections System Manifold Size on Flame Protection Equipment. Paper presented to various petrochemical companies, Westech Industrial Limited, Calgary, Alberta, Canada.

Lapp, K. and Werneburg, H. 1991. Recent Research Advancing the Development of Standards for Detonation Flame Arresters. Paper presented at the Marichem '91 Conference, December 3–5, 1991, Cologne, Germany.

Lapp, K. and Werneburg, H. 1993. Detonation Flame Arrester Qualifying Application Parameters for Explosion Prevention in Drilling and Production. Paper presented at the Calgary/Code/Cadoc Spring Drilling Conference, May 4, 1993, Westech Industrial Limited, Calgary, Alberta, Canada.

Lapp, K. and Werneburg, H. 1999. Detonation Flame Arrester Application Parameters for Explosion Prevention in Vapour Handling Systems. *J. Canadian Petroleum Technology, Special Edition, 1949-1999,* Paper 94–58.

Lightfoot, F. K. 1977. Flame Arresters-Insure Their Protection. *Professional Safety,* February, pp. 44–47.

Mendoza, V. A., Smolensky, V. G., and Straitz, J. F. III. 1998. Do Your Flame Arresters Provide Adequate Protection? *Hydrocarbon Process., Intl. Ed.,* 77(10), 63–64, 66, 69.

Quinton, P. G. 1962. The Flow Resistance of Flame Arresters. *British Chem. Engng.,* 9(12), 914–921.

Roussakis, N. 1988. The Hazard of an Oil Storage Tank Explosion from Lightning Strike at a Flare Stack. Technical Paper (May1988). Westech Industrial Limited, Calgary, Alberta, Canada

Roussakis, N. and Lapp. K. 1989. Prevention of Oil Storage Tank Explosions in Flaring Operations. Paper presented at the Canadian Petroleum Association Accident Prevention Committee, March 20, 1989, Calgary, Alberta, Canada.

Roslovskii, A. I. And Zakaznov, V. F. 1971. Effect of Gas Expansion During Combustion on the Possibility of Using Flame Arresters. *Combustion and Flame,* 17, 215–221.

Seebold, J. G. 1992. Detonation Arresters for Marine Vapor Control Systems: Status of Research and Uncertainties—Where from Here? Paper presented at a Colloquium on Vapor Control Systems: Detonation Arrester Research and Testing, July 23, 1992, Health and Safety Executive, Buxton, Derbyshire, England.

Senecal, J. A. and Meltzer, J. S. 1992. Barrier Detonation Arresting Systems. Paper presented at the API Marine Technical–Environmental Conference, January 8–9, 1992, Chantilly, VA.

Thomas. G. O. and Teodorczyk. 1998. An Evaluation of New Procedures for Testing Explosion Arresters. *Trans. IChemE,* 76, Part B, 277–285 (November).

Vickers, K. 1993. Detonation Arresters for Kraft Pulp Mills for Use in NCG Collection Systems. Paper presented to various pulp mills. Westech Industrial Limited, Calgary, Alberta, Canada.

Vickers, K. and Knittel, T. no date. Detonation Arresters for Offshore Production Platforms. Paper presented to API Task Force for API RP 14C. Westech Industrial Limited, Calgary, Alberta, Canada.

Watson, P. B. 1988. Flame Arresters and the Chemical Tanker. *IChemE Loss Prevention Bulletin* No. 079 (February 1988).

Werneburg, H. 1992. Test Instrumentation and Standardization of Data Collection Methods. Paper presented at the International Data Exchange Symposium, October 27–29, Banff, Alberta, Canada.

Werneburg, H. and Lapp, K. 1997. Detonation Arrester Testing and Selection: A North American Perspective. Paper presented at the CONCAWE Safety Seminar on Vapour Recovery Units, March 5, 1997, Brussels, Belgium.

Glossary

Autoignition Temperature (AIT): The autoignition temperature of a substance, whether solid, liquid, or gaseous, is the minimum temperature required to initiate or cause self-sustained combustion, in air, with no other source of ignition.

Bidirectional Flame Arrester: An in-line flame arrester is considered to be bidirectional if:

(a) the investigation (test) shows that the flame arrester performs its intended function with a detonation or deflagration approaching from either direction, or

(b) the design of the flame arrester is symmetrical and each end is considered identical when approached by a detonation or deflagration from either direction.

Burnback Resistance: The ability of a flame arrester to maintain its mechanical integrity and prevent flame breakthrough when the flame arrester is subjected to a standing flame. See also Endurance Burn(ing).

Burning Velocity: Burning velocity is the speed at which a flame front propagates relative to the unburned gas. This differs from flame speed. The laminar burning velocity (S_L) is the speed at which a laminar (planar) combustion wave propagates relative to the unburned gas mixture ahead of it. The fundamental burning velocity (S_u) is similar, but generally not identical to, the observed laminar burning velocity. This is because S_u is a characteristic parameter referring to standardized unburned gas conditions (normally 760 mm Hg and 25°C), and which has been corrected for nonidealities in the measurement. The turbulent burning velocity (S_t) exceeds the burning velocity measured under laminar conditions to a degree depending on the scale and intensity of turbulence in the unburned gas.

Calculated Adiabatic Reaction Temperature (CART): See Flame Temperature.

Cascading: See Pressure Piling

Chemical Isolation: The stopping of a flame by means of a chemical suppressant.

Combustible Liquid: A term used to classify certain liquids that will burn on the basis of flash points. The National Fire Protection Association (NFPA) defines a combustible liquid as any liquid that has a closed-cup flash point above 100°F (37.8°C) (NFPA 30). There are three subclasses, as follows:

Class II liquids have flash points at or above 100°F (37.8°C) but below 140°F (60°C)

Class III liquids are subdivided into two additional subclasses:

Class IIIA: Those having flash points at or above 140°F (60°C) but below 200°F (93.4°C).

Class IIIB: Those having flash points at or above 200°F (93.4°C).

The Department of Transportation (DOT) defines "combustible liquids" as those having flash points of not more than 141°F (60.5°C) and below 200°F (93.4°C).

Combustible Concentration Reduction: The technique of maintaining the concentration of combustible material in a closed space below the lower flammable limit.

Combustion: A chemical process of oxidation that occurs at a rate fast enough to produce heat and usually light, in the form of either a glow or flames.

Confined Volume Deflagration: A deflagration occurring in an enclosed volume such as a pipe or pressure vessel.

Conservation Vent Valve: A device designed to maintain pressure within preset limits in a liquid-containing vessel for the purpose of emissions reduction. It also provides protection against excessive pressure or vacuum.

Continuous Flame Test: A test in which a flame arrester is subjected to flame of a continuously burning mixture (as specified in UL 525 for deflagration or detonation flame arresters) on the outlet face of the arrester for one hour (or longer at the manufacturer's request).

Crimped Metal Ribbon: A flame arrester element that is manufactured of alternate layers of thin corrugated metal ribbon and a flat metal ribbon that are wound together on a mandrel to form a cylindrical assembly of many layers to produce a range of different sized triangular cells. The height and width of the triangular cells can be varied to provide the required quenching diameter.

Critical Diameter: See Quenching Diameter.

DDA: Deflagration and detonation arresters or deflagration and detonation flame arresters.

Decomposition Flames: Flames that are produced by exothermic decomposition of certain gases in the absence of any oxidant, provided that they are above minimum conditions of pressure, temperature, and pipe diameter. Common examples include acetylene, ethylene oxide, and ethylene.

Deflagration: A combustion wave (flame) propagating in the unreacted medium at a velocity that is less than the speed of sound as measured at the flame front, and which propagates via a process of heat transfer and species diffusion across the flame front. The flame front is not coupled in time nor space with the preceding weak shock front. The pressure peak coincides with the flame front but a marked pressure rise precedes it. Typical maximum pressure ratios generated by deflagrations in confined volumes are in the range of 8-12. Maximum propagation velocities of 10-100 m/s are typical, although up to several hundred m/s may be observed.

Deflagration Flame Arrester: A flame arrester used to prevent the transmission of a deflagration.

Deflagration Isolation: A method employing equipment and procedures that interrupts the propagation of a deflagration flame front past a point (usually in a pipe).

Deflagration Pressure Containment: The technique of specifying the design pressure of a vessel and its appurtenances so that they are capable of withstanding the maximum pressures resulting from an internal deflagration.

Deflagration Suppression: The technique of detecting and arresting combustion in a confined space while the combustion is still in its incipient stage, thus preventing the development of pressures that could result in an explosion.

Deflagration-to-Detonation Transition (DDT): The transition phenomenon resulting from the acceleration of a deflagration flame to detonation via flame-generated turbulent flow and compressive heating effects. At the instant of transition a volume of precompressed, turbulent gas ahead of the flame front detonates at unusually high velocity and overpressure.

Deflagration Venting: The reduction of pressure generated in a vessel by a deflagration by allowing the emergency flow of the vessel contents from the vessel by means of an opening in the vessel, thus avoiding the failure of the vessel by overpressure. The vent opening is usually closed by a pressure-relieving cover (e.g., rupture disk, explosion disk or hatch).

Detonation: Propagation of a flame-driven shock wave at a velocity at or above the speed of sound in the unreacted medium as measured at the flame front. The wave is sustained by chemical energy released by shock compression and ignition of the unreacted medium. The flame front is coupled in time and space with the shock front, and there is no pressure increase significantly ahead of the shock-flame front. Propagation velocities in the range 1000–3500 m/s may be observed depending on the gas mixture, initial temperature and pressure, and type of detonation.

Detonation Flame Arrester: A flame arrester used to prevent the transmission of a detonation.

Detonable Limits: The minimum and maximum concentrations of a combustible material, in a homogeneous mixture with a gaseous oxidizer, that will propagate a detonation.

Detonation Momentum Attenuator: A mechanical device inside of a detonation flame arrester whose purpose is to reduce both the high pressure and the dynamic energy of a detonation and to split the flame front before it reaches the actual flame arrester element, thus avoiding structural damage to the element. (This device is also called a "shock absorber" device by some manufacturers.)

Dry Type Flame Arrester: A flame arrester that uses an element consisting of small gaps or apertures to quench and extinguish the flame, as distinguished from hydraulic flame arresters.

Dust: Any finely divided solid, 420 microns or 0.017 inches, or less in diameter (that is, material that can pass through a U.S. No. 40 standard sieve).

End-of-Line Flame Arrester: A deflagration flame arrester that is installed on atmospheric pressure tank vent nozzles, having only one pipe connection. It is also called a tank vent deflagration flame arrester.

Endurance Burn(ing): Steady burning of a stabilized flame at or close to the flame arrester element.

Endurance Burn Test: A test in which a flame arrester is subjected to a stable flame on the surface of the arrester for a length of time (depends on whether UL 525 or the USCG protocol is used) until the highest obtainable temperature is reached on the ignited side or until the temperature on the protected side has a temperature rise of 100°C.

Equivalence Ratio: The ratio of fuel concentration in the actual fuel–air mixture divided by the fuel concentration in a stoichiometric mixture.

Explosion: The bursting or rupturing of an enclosure or container due to the development of internal pressure from a deflagration or detonation.

Failure: An unacceptable difference between expected and observed performance.

Fast-Acting Valve: A valve that closes a path of deflagration propagation in a pipe or duct in response to upstream detection of a deflagration.

Flame: A region in which chemical interaction between gases occurs, accompanied by the evolution of light and heat (see Decomposition Flames).

Flame Arrester: A device fitted to the opening of an enclosure or to the connecting piping of a system of enclosures and whose intended function is to allow flow but prevent the transmission of flame from either a deflagration or detonation.

Flame Arrester Element (Matrix): That portion of a flame arrester whose principal function is to prevent flame transmission, usually by quenching the flame front.

Flame Arrester Housing: That portion of flame arrester whose principal function is to provide a suitable enclosure for the flame arrester element, and to facilitate mechanical connection to other systems.

Flame Barrier: A device that prevents transmission of a flame from a source to a receptor.

Flame Front: That portion of the flame reaction zone moving into the unburned gas where the bulk of the reaction occurs and the medium reaches its ignition temperature.

Flame Front Diverter: A device that opens in response to the pressure wave preceding the flame front of the deflagration, venting the flame front and pressure wave.

Flame Propagation: The movement of a flame front in piping or equipment.

Flame Speed: The speed of a flame front relative to a fixed reference point. Flame speed is dependent on turbulence, the equipment geometry, and the fundamental burning velocity.

Flame Temperature: Theoretical temperature achieved based on chemical equilibrium with the assumption of Gibbs free energy minimization

Flame Trap: Another name for a flame arrester, commonly used in the United Kingdom.

Flame Velocity: See Flame Speed.

Flammable Limits: The minimum and maximum concentration of fuel vapor or gas in a fuel vapor or gas/gaseous oxidant mixture (usually expressed in percent by volume) defining the concentration range (flammable or explosive range) over which propagation of flame will occur on contact with an ignition source. See also Lower Flammable Limit and Upper Flammable Limit.

Flammable Liquid: A term used to classify certain liquids as defined by NFPA 30 as a liquid with a closed-cup flash point below 100°F (37.8°C) and Reid vapor pressures not exceeding 40 psia at 100°F (37.8°C). Flammable liquids are called Class I liquids and have three subdivisions as follows:

Class IA: Those having flash points below 73°F (22.8°C) and boiling points below 100°F (37.8°C).

Class IB: Those having flash points below 73°F (22.8°C) and boiling points at or above 100°F (37.8°C).

Class IC: Those having flash points at or above 73°F (22.8°C) but below 100°F (37.8°C).

Flammable Range: The range of concentrations between the lower and upper flammability limits.

Flashback: Undesired flame propagation opposite to the direction of flow. It is also used to describe failure of a flame arrester element.

Flow Controlled Aperture: An aperture designed to produce flow velocities which exceed the local flame speed of the flammable mixture, thus preventing flame transmission in the reverse direction.

Fundamental Burning Velocity: See Burning Velocity

Galloping Detonation: A detonation that periodically fails and reinitiates during propagation. This type of detonation is typically observed in near-limit mixtures. Since it reinitiates via DDT, a galloping detonation is periodically overdriven and results in large overpressures at periodic distances along a pipe.

Gas: The state of matter characterized by complete molecular mobility and unlimited expansion at standard temperature and pressure.

Gas Enrichment: The addition of another flammable gas to a gas–air mixture to raise the concentration above the upper flammable limit.

Hazard: An inherent chemical or physical characteristic that has the potential for causing damage to people, property, or the environment. In this book, it is typically the combination of a hazardous material, an operating environment, and certain unplanned events that could result in an accident.

Hazardous Material: In a broad sense, any substance or mixture of substances having properties capable of producing adverse effects to the health or safety of human beings or the environment. Material presenting dangers beyond the fire problems relating to flash point and boiling point. These dangers may arise from, but are not limited to, toxicity, reactivity, instability, or corrosivity.

High Velocity Vent Valve: A device to prevent the passage of flame in the reverse direction, consisting of a mechanical valve which adjusts the opening available for flow in accordance with the pressure at the inlet of the valve in such a way that the efflux (exiting) velocity cannot be less than 30 m/s.

Hybrid Mixture: A mixture of a combustible gas with either a combustible dust or combustible mist.

Hydraulic Diameter: An equivalent diameter for noncircular apertures which is equal to 4× aperture area/aperture perimeter.

Hydraulic Flame Arrester: A flame arrester consisting of a vessel filled with a seal fluid (often water) and a distributor which breaks up the incoming gas into discrete bubbles, thus facilitating quenching of the flame and preventing flame transmission.

Inert Gas: A nonflammable, nonreactive gas that can be used to render the combustible material in a system incapable of supporting combustion.

Inerting: A technique by which a combustible mixture is rendered nonignitable by addition of an inert gas or a noncombustible dust.

Initial Pressure: The maximum normal operating pressure that exists at a point in the system prior to initiation of a flame event.

In-Line Flame Arrester: A flame arrester which is fitted with two pipe connections, one on each side of the flame arrester element. An in-line flame arrester may be either a deflagration or a detonation flame arrester. An in-line arrester that is installed on a vessel vent nozzle with a short length of pipe attached to the discharge (atmospheric) side is sometimes also called a "pipe-away" flame arrester.

Interlock System: A system that detects out-of-limits or abnormal conditions or improper sequences and either halts further action or starts corrective action.

Isolation: A means of preventing certain stream properties (deflagration, mass flow, ignition capability) from being conveyed past a predefined point.

Laminar Burning Velocity: See Burning Velocity

Limiting Oxidant Concentration (LOC): The concentration of oxidant below which a deflagration cannot occur in a specified mixture.

Liquid Seal: A device for preventing the passage of flame by passing the gas mixture through a suitable liquid. See Hydraulic Flame Arrester.

Lower Flammable Limit (LFL): That concentration of a combustible material in air below which ignition will not occur. It is often, interchangeably, called Lower Explosive Limit (LEL). Mixtures below this limit are said to be "too lean."

Material Choke: A mass of bulk solids or powders in a rotary valve or screw feeder that prevents a flame from being transmitted.

Maximum Experimental Safe Gap (MESG): The maximum gap of the joint between the two parts of the interior chamber of a test apparatus which, when the internal gas mixture is ignited and under specified conditions, prevents ignition of the internal gas mixture through a 25-mm-long joint, for all concentrations of the tested gas or vapor in air. The MESG is a property of the respective gas mixture, but can vary depending on the test apparatus.

Minimum Igniting Current (MIC) Ratio: The ratio of the minimum current required from an inductive spark discharge to ignite the most easily ignitable mixture of a gas or vapor, divided by the minimum current required from an inductive spark discharge to ignite methane under the same test conditions.

Minimum Ignition Energy (MIE): The minimum amount of energy released at a point in a combustible mixture that caused flame propagation away from the point, under specified test conditions. The lowest value of the minimum ignition energy is found at a certain optimum mixture. The lowest value is usually quoted as the minimum ignition energy.

Mist: A dispersion of fine liquid droplets in a gaseous medium.

Mitigation: Lessening the risk of an accident event sequence by acting on the source in a preventive way by reducing the likelihood of occurrence of the event, or in a protective way by reducing the magnitude of the event and/or the exposure of local persons or property.

Operating Pressure: The maximum pressure at which a flame arrester can be used according to its certification.

Optimum Mixture: A specific mixture of fuel and oxidant that yields the most rapid combustion at a specific measured quantity or that yields the lowest value of the minimum ignition energy or that produces the maximum deflagration pressure. The optimum mixture is not always the same for each combustion property that is measured.

Overdriven Detonation: The unstable condition that exists during a deflagration-to-detonation transition (DDT) before a state of stable detonation is reached. Transition occurs over the length of a few pipe diameters and propagation velocities of up to 2000 m/s have been measured for hydrocarbons in air. This is greater than the speed of sound as measured at the flame front. Overdriven detonations are typically accompanied by side-on pressure ratios (at the pipe wall) in the range 50–100. A severe test for detonation flame arresters is to adjust the run-up distance so the DDT occurs at the flame arrester, subjecting the device to the overdriven detonation impulse.

Overpressure: The pressure increase above normal operating pressure that is caused by the flame.

Oxidant: Any gaseous material that can react with a fuel (either gas, dust, or mist) to produce combustion. Oxygen in air is the most common oxidant.

Oxidant Concentration Reduction: The technique of maintaining the concentration of the oxidant in a closed space below the concentration required for combustion to occur.

Porous Media: Any solid material through which gas can flow, and that affords some measure of heat extractive capability and flame quenching by cooling.

Pressure Piling: In a compartmented system in which there are separate but interconnected volumes, the pressure developed by the deflagration in one compartment causes a pressure rise in the unburned gas in the interconnected compartment, so that the elevated pressure in the latter compartment becomes the starting pressure for a further deflagration. This effect is known as pressure piling, or cascading.

Pre-Volume Vessel: A process vessel or piping system (length to diameter ratio of less than 5) in which a confined deflagration occurs (as defined in CEN Standard EN 12874).

Process Hazard Analysis (PHA): The identification of undesired events that lead to hazardous events, the analysis of the mechanisms by which these undesired events could occur, and usually the estimation of the consequences.

Protected Side of Flame Arrester: The side of a flame arrester and system to which it is connected where flame is to be excluded.

Purge Gas: A gas that is continuously or intermittently added to a system to render the atmosphere noncombustible. The purge gas can be inert or combustible.

Quenching: Rapid cooling of a flame in a short time (almost instantaneously) to extinguish it.

Quenching Diameter: The largest diameter of a cylindrical tube that will just quench (extinguish) the flame front of a particular fuel–air mixture.

Quenching Distance: The distance between two parallel plates (flat walls) that will just quench (extinguish) the flame front of a particular fuel-oxidant mixture. It is smaller than the quenching diameter in cylindrical tubes.

Quenching Length: The length of a flame arrester element necessary to quench a high speed flame.

Run-up Distance or Run-up Length: The distance in the direction of flame propagation from the point of ignition to any point in a pipe system. Deflagration flames accelerate over this distance due to turbulence and precompression effects. Depending on pipe diameter, surface roughness, and the presence of turbulence-producing obstacles (elbows, valves, etc.) this distance may be sufficient for DDT to occur.

Safe Burning Time: The period of stabilized burning on a flame arrester without flame being transmitted through the arrester.

Sensitive Gas: A gas that is much more likely to cause a failure in a flame arrester test because of its low AIT or other characteristics not obvious from its IEC standard MESG value.

Shock Absorber Device: See Detonation Momentum Attenuator.

Shock Wave: A transient change in the gas density, pressure, and velocity of the air surrounding an explosion point. The initial change can be either discontinuous or gradual. A discontinuous change is referred to as a shock wave, and a gradual change is known as a pressure wave.

Specific Impulse: The area under the overpressure versus time curve.

Stable Detonation: A detonation that progresses through a confined system without significant variation of velocity and pressure characteristics. For atmospheric conditions, typical velocities range between 1600 and 2200 m/s for standard test mixtures and test procedures.

Stabilized Burning: Steady burning of a flame, stabilized at, or close to the flame arrester element.

Static Flame Arrester: A flame arrester designed to prevent flame transmission by quenching gaps (apertures). These are usually dry type flame arresters with elements such as crimped metal ribbon, parallel plates, wire gauze (mesh), and sintered metal.

Stoichiometric Mixture: A balanced mixture of fuel and oxidizer such that no excess of either remains after combustion.

Suppressant: A chemical agent used in a deflagration suppressant system to extinguish the deflagration.

Suppressant Barrier: An isolation system using a suppressant.

Turbulent Burning Velocity: See Burning Velocity.

Unidirectional Flame Arrester: An in-line flame arrester that is designed to stop flame propagation approaching from only one direction.

Unstable Detonation: See Overdriven Detonation.

Unprotected Side of Flame Arrester: The side of a flame arrester and system to which it is connected where flame is expected

Upper Flammable Limit (UFL): That concentration of a combustible material in air above which ignition will not occur. It is often, interchangeably, called Upper Explosive Limit (UEL). Mixtures above this limit are said to be "too rich."

Vapor: A gas that is at a temperature below the critical temperature, so that it can be liquefied by compression, without lowering its temperature. It is the gas phase of a substance, particularly of those that are normally liquids or solids at ordinary temperatures.

Vapor Collection System: A piping system to which vessels are connected, that collects vapors from these vessels and directs them to environmental control equipment such as flares, incinerators, scrubbers, and activated carbon adsorbers.

Velocity Flame Stopper: See Flow Controlled Aperture.

Vent: An opening for the passage of, or dissipation of, fluids, such as gases, fumes, smoke, and the like.

Vent Manifold: See Vapor Collection System.

Ventilation: The process of supplying or removing an atmosphere to or from any space by natural or mechanical means.

Venting: See "Deflagration Venting."

Index

G

H

I

P

Q

R

S

T

U

V

W

Z

Publications Available from the CENTER FOR CHEMICAL PROCESS SAFETY of the AMERICAN INSTITUTE OF CHEMICAL ENGINEERS 3 Park Avenue, New York, NY 10016-5991

CCPS Guidelines Series

Guidelines for Process Safety in Outsourced Manufacturing Operations
Guidelines for Process Safety in Batch Reaction Systems
Guidelines for Chemical Process Quantitative Risk Analysis, 2nd Edition
Guidelines for Consequence Analysis of Chemical Releases
Guidelines for Pressure Relief and Effluent Handling Systems
Guidelines for Design Solutions for Process Equipment Failures
Guidelines for Safe Warehousing of Chemicals
Guidelines for Postrelease Mitigation in the Chemical Process Industry
Guidelines for Integrating Process Safety Management, Environment, Safety, Health, and Quality
Guidelines for Use of Vapor Cloud Dispersion Models, Second Edition
Guidelines for Evaluating Process Plant Buildings for External Explosions and Fires
Guidelines for Writing Effective Operations and Maintenance Procedures
Guidelines for Chemical Transportation Risk Analysis
Guidelines for Safe Storage and Handling of Reactive Materials
Guidelines for Technical Planning for On-Site Emergencies
Guidelines for Process Safety Documentation
Guidelines for Safe Process Operations and Maintenance
Guidelines for Process Safety Fundamentals in General Plant Operations
Guidelines for Chemical Reactivity Evaluation and Application to Process Design
Tools for Making Acute Risk Decisions with Chemical Process Safety Applications
Guidelines for Preventing Human Error in Process Safety
Guidelines for Evaluating the Characteristics of Vapor Cloud Explosions, Flash Fires, and BLEVEs
Guidelines for Implementing Process Safety Management Systems
Guidelines for Safe Automation of Chemical Processes
Guidelines for Engineering Design for Process Safety
Guidelines for Auditing Process Safety Management Systems
Guidelines for Investigating Chemical Process Incidents
Guidelines for Hazard Evaluation Procedures, Second Edition with Worked Examples
Plant Guidelines for Technical Management of Chemical Process Safety, Revised Edition
Guidelines for Technical Management of Chemical Process Safety
Guidelines for Process Equipment Reliability Data with Data Tables
Guidelines for Safe Storage and Handling of High Toxic Hazard Materials
Guidelines for Vapor Release Mitigation

CCPS Concepts Series

Deflagration and Detonation Flame Arresters
Making EHS an Integral Part of Process Design
Revalidating Process Hazard Analyses
Electrostatic Ignitions of Fires and Explosions
Evaluating Process Safety in the Chemical Industry
Avoiding Static Ignition Hazards in Chemical Operations
Estimating the Flammable Mass of a Vapor Cloud
RELEASE: A Model with Data to Predict Aerosol Rainout in Accidental Releases
Practical Compliance with the EPA Risk Management Program
Local Emergency Planning Committee Guidebook: Understanding the EPA Risk Management Program Rule
Inherently Safer Chemical Processes: A Life-Cycle Approach
Contractor and Client Relations to Assure Process Safety
Understanding Atmospheric Dispersion of Accidental Releases
Expert Systems in Process Safety
Concentration Fluctuations and Averaging Time in Vapor Clouds

Proceedings and Other Publications

Center for Chemical Process Safety International Conference and Workshop: Making Process Safety Pay—The Business Case
Center for Chemical Process Safety International Conference and Workshop: Process Industry Incidents—Investigation Protocols, Case Histories, Lessons Learned, 2000
Proceedings of the International Conference and Workshop on Modeling the Consequences of Accidental Releases of Hazardous Materials, 1999
Proceedings of the International Conference and Workshop on Reliability and Risk Management, 1998
Proceedings of the International Conference and Workshop on Risk Analysis in Process Safety, 1997
Proceedings of the International Conference and Workshop on Process Safety Management and Inherently Safer Processes, 1996
Proceedings of the International Conference and Workshop on Modeling and Mitigating the Consequences of Accidental Releases of Hazardous Materials, 1995
Proceedings of the International Symposium and Workshop on Safe Chemical Process Automation, 1994
Proceedings of the International Process Safety Management Conference and Workshop, 1993
Proceedings of the International Conference on Hazard Identification and Risk Analysis, Human Factors, and Human Reliability in Process Safety, 1992
Proceedings of the International Conference and Workshop on Modeling and Mitigating the Consequences of Accidental Releases of Hazardous Materials, 1991
Safety, Health and Loss Prevention in Chemical Processes: Problems for Undergraduate Engineering Curricula

Printed and bound by CPI Group (UK) Ltd, Croydon, CR0 4YY

Printed and bound by CPI Group (UK) Ltd, Croydon, CR0 4YY
16/09/2026
14963420-0003